D1548234

WITHDRAWN

WORKPLACE HEALTH PROTECTION

Industrial

Hygiene

Program

Guide

by Robert G. Confer
with support from
Mobil Oil Corporation

WORKPLACE HEALTH PROTECTION

Industrial Hygiene Program Guide

LEWIS PUBLISHERS
Boca Raton Ann Arbor London Tokyo

Library of Congress Cataloging-in-Publication Data

Confer, Robert G.
Workplace health protection : industrial hygiene program guide / by Robert G. Confer.
p. cm.
ISBN 0-87371-387-7
1. Industrial hygiene--Standards. 2. Industrial hygiene--Evaluation. I. Title
RC967.C5967 1994
363.11'0218—dc20 93-45604
CIP

Direct all inquiries to CRC Press, Inc., 2000 Corporate Blvd., N.W., Boca Raton, Florida 33431.

Lewis Publishers is an imprint of CRC Press

International Standard Book Number 0-87371-387-7
Library of Congress Card Number 93-45604
Printed in the United States of America 1 2 3 4 5 6 7 8 9 0
Printed on acid-free paper

FOREWORD

Criteria to evaluate the effectiveness of a comprehensive industrial hygiene program were included in a publication issued by AIHA in 1976. The document, titled "Standards, Interpretations and Audit Criteria for Performance of Occupational Health Programs", was designed to help make occupational health programs more effective. Audit questionnaires were included to aid in assessing and reviewing the operation of occupational health programs covering such elements as administration, medical, nursing, industrial hygiene, and health physics. To my knowledge, no definitive work has been published since, although a number of companies have in-house criteria. Many of these have a regulatory compliance focus.

The need for up-to-date and modern criteria for conducting reviews of comprehensive industrial hygiene programs has become apparent to many industrial hygienists, particularly in overseas operations. Typically, overseas programs have not developed as rapidly as U.S. counterparts, where regulations often have required specific actions for which compliance can be readily determined. As a result, personnel with responsibility for overseas programs may not be as familiar with up-to-date criteria for evaluating industrial hygiene programs. Additionally, program reviews at the same facility by different industrial hygienists in succeeding years often can produce conflicting results. One individual may determine that a program element, such as that for hearing conservation or respiratory protection, is effective one year and another person may find deficiencies in a succeeding year.

During the last few years of my employment, I started to formulate performance criteria for evaluating industrial hygiene program elements, which eventually reached about 50 elements. I had intended to develop these for publication, but time did not permit completing the task. Shortly after joining Mobil Oil Corporation as a consultant, I was requested to develop a Workplace Health Protection Program Reference Document for worldwide use. The main objective was to provide a PRACTICAL TOOL for evaluating the applicability and/or effectiveness of workplace health protection programs on a CONSISTENT basis in upstream and downstream petroleum operations, petrochemical facilities, and other company operations, both domestically and overseas. This is the result of that effort.

Robert G. Confer

The Author

Robert Confer has worked in the field of industrial hygiene for over 36 years. Following several years of field work as an industrial hygienist with the Pennsylvania Department of Health, he worked at the Bettis Atomic Power Laboratory for two years. From there he went to Exxon Corporation, from which he retired in 1989 after 25 years of service as an industrial hygienist with various affiliated companies. Primary assignments were as a field industrial hygienist and conducting research on industrial hygiene instrumentation and sampling method development. Following his retirement he worked as a private contractor with the Employee and Facility Safety Division of Mobil Oil Corporation.

Work assignments during his career have taken him from coal, clay, molybdenum and uranium mines to petroleum, natural gas, marine transportation, petrochemical, semiconductor, and general manufacturing operations in more than 30 states and 25 foreign countries. He is the author of 27 articles that have been published in peer reviewed journals, has contributed chapters to several books, and has been awarded three U.S. patents. He continues to be active in the field.

TABLE OF CONTENTS

1. INTRODUCTION

INTRODUCTION

A comprehensive industrial hygiene program review is very broad in scope and difficult to complete in even a small facility without approaching it in an organized manner. Sufficient time is seldom available while working in the field to develop a review procedure that could be applied uniformly over time, by different individuals, or even the same person, that enables industrial hygiene program assessments on a consistent basis. Regulatory audits do not necessarily accomplish this since they are often narrow in scope, and do not consider all aspects of what can be referred to generically as a comprehensive workplace health protection program.

A comprehensive workplace health protection program incorporates a large number of separate but interrelated elements which address the potential health concerns associated with the operations being carried out. This is necessary to prevent adverse health effects or significant discomfort among employees, or residents in the surrounding community. In order to ensure its continued effectiveness, a program should be reviewed periodically, deficiencies identified, and measures developed and implemented promptly for improvement. Such reviews must be carried out using a procedure that provides a consistent basis for program assessment.

This Workplace Health Protection Program Reference Document provides such a tool. Since it does not address regulatory issues from any jurisdiction, it can be applied to large and small facilities throughout the world. The information presented for each of the fifty elements of this reference document is based on established industry practice, consensus standards, the technical literature, trade association publications, and procedures that have evolved over time and been implemented based on their effectiveness in protecting the health of employees.

A comprehensive program for some facilities could include all of the elements incorporated here. Others may find that only a few elements are applicable to their operation. Determining which apply is the first step in conducting a program review.

Table 1 in the Introduction is a starting point for carrying out a workplace health protection program assessment using this tool. The reviewer is to read the brief statement relative to each element and respond to the questions related to that program element's status. This will provide a basis for judging whether the element is needed, whether it is in place, and if so, if it is acceptable. If in place and not acceptable, or not in place, a notation is made as to the priority for its development and implementation. If a judgment cannot be made regarding whether the element is needed, in place, or acceptable, then a more detailed assessment should be carried out. This is accomplished by referring to the specific discussion of the topic that is incorporated into one of eight program categories and responding to the questions related to that element.

Each of the elements included in this program reference document contains an objective statement, general background information on the topic discussed, a considerations section addressing exposure/control issues, and a checklist. The checklist is employed as the means for determining, in detail, the effectiveness of the program element in meeting facility needs. The questions are phrased to elicit positive replies when acceptable. Thus, a "No", or "Don't Know" answer may indicate a deficiency with respect to the issue addressed in the question. Corrective action that may be needed is typically based on the questions for which there is a "No" response. Additional information will need to be obtained before a "Don't Know" response can be effectively answered with a yes or no reply.

The completed element checklists provide an indication of the status of the total workplace health protection program's effectiveness for meeting facility needs. The information contained in them can be used as a basis for seeking management support for program improvement, prioritizing work, redirecting effort, obtaining additional resources, as well as identifying other measures that need to be developed and implemented for program improvement.

TABLE 1

STATUS OF WORKPLACE HEALTH PROTECTION PROGRAM ELEMENTS

ELEMENT	CONSIDERATIONS	PROGRAM ELEMENT: NEEDED* Y/N	IN PLACE* Y/N	ACCEPTABLE** Y/N	PRIORITY*** H/M/L
MANAGEMENT					
Administration	Personnel with responsibility for workplace health protection activities should have clearly defined objectives, be accountable to management for accomplishing these, and provide support in developing, implementing, and maintaining an effective workplace health protection program.				
Comment					
Communications	Communications among industrial hygiene, management, safety, employee relations, medical, operations, maintenance, technical, and other groups should be ongoing and effective. The industrial hygiene (workplace health protection) function should be represented on appropriate committees to maintain awareness of workplace health issues.				
Comment					

* Based on Judgement: Y-Yes; N-No.

** Based on Judgement with Reference to the Information Presented Under the Considerations Heading: Y-Yes; N-No.

*** Based on Judgement for Development/Implementation: H-High; M-Medium; L-Low.

ELEMENT	CONSIDERATIONS	NEEDED* Y/N	IN PLACE* Y/N	ACCEPTABLE** Y/N	PRIORITY*** H/M/L
		PROGRAM ELEMENT			
MANAGEMENT (Continued)					
Planning/Stewardship	Where appropriate, a workplace health protection action plan should be developed annually by the industrial hygiene function, submitted to management for review and approval, and stewarded as per management direction. This procedure should be a routine part of the industrial hygiene program administrative function.				
Comment					
Resources	Management will determine the required time and resources to be devoted to the workplace health protection activities to accomplish planned program objectives, meet regulatory requirements, and respond to requests/complaints/concerns. The resources required include personnel; funding and equipment; reference literature; a written plan that has been reviewed and approved by management; and qualitative exposure assessment information.				
Comment					

* Based on Judgement: Y-Yes; N-No.
** Based on Judgement with Reference to the Information Presented under the Considerations Heading: Y-Yes; N-No.
*** Based on Judgement for Development/Implementation: H-High; M-Medium; L-Low.

ELEMENT	CONSIDERATIONS	NEEDED* Y/N	IN PLACE* Y/N	ACCEPTABLE** Y/N	PRIORITY*** H/M/L
TECHNICAL ASPECTS					
Continuous Monitors	Where appropriate, continuous monitors are operated to provide early warning of contaminant releases and potentially hazardous exposure situations. To be effective, the instruments must be selected for the environment in which they will be operated, placed in the right locations, be properly maintained, and periodically checked (for operational response) and calibrated, as necessary to ensure proper operation. Records should be retained of the response checks and calibrations.				
Comment	______				
Ergonomics	Some tasks could be done more effectively by redesigning the job to fit the person rather than forcing the person to fit the job. The workplace must be assessed to identify and correct potential hazards from repetitive motion, improper tool use, improper tool design, etc. Review of operations such as welding, hand tool use, equipment location, and methods of performing a task can provide information which can facilitate changes for reducing glare, cumulative trauma disorders, strains, and other adverse effects on employees.				
Comment	______				

PROGRAM ELEMENT: NEEDED, IN PLACE, ACCEPTABLE, PRIORITY

* Based on Judgement: Y-Yes; N-No.

** Based on Judgement with Reference to the Information Presented Under the Considerations Heading: Y-Yes; N-No.

*** Based on Judgement for Development/Implementation: H-High; M-Medium; L-Low.

ELEMENT	CONSIDERATIONS	PROGRAM ELEMENT NEEDED* Y/N	IN PLACE* Y/N	ACCEPTABLE** Y/N	PRIORITY*** H/M/L
TECHNICAL ASPECTS (Continued)					
Hearing Conservation	Where there is a potential for hearing damage to occur, a comprehensive hearing conservation program should be implemented. This involves noise exposure assessment; development of facility noise contour maps; identification of personnel to include in the hearing conservation program; employee training; posting of high noise areas; identification of noise sources amenable to engineering control; selecting and providing personal hearing protection; audiometric testing of noise exposed personnel using appropriate equipment (audiometer/booth); and the periodic assessment of audiometric results to determine the effectiveness of the program.				
Comment					
Heat Stress	Employee exposure to heat stress situations can occur in petroleum and petrochemical process operations. These should be identified, and personnel who may be at risk should be made aware of the exposure hazard and potential adverse health effects. Exposures should be assessed and controls implemented, as appropriate. A discussion of heat stress effects should be included in the employee hazard awareness training program.				
Comment					

* Based on Judgement: Y-Yes; N-No.

** Based on Judgement with Reference to the Information Presented Under the Considerations Heading: Y-Yes; N-No.

*** Based on Judgement for Development/Implementation: H-High; M-Medium; L-Low.

ELEMENT	CONSIDERATIONS	PROGRAM ELEMENT NEEDED* Y/N	IN PLACE* Y/N	ACCEPTABLE** Y/N	PRIORITY*** H/M/L
TECHNICAL ASPECTS (Continued)					
Ionizing Radiation	Sources of ionizing radiation often are used within a facility for X-raying welds, determining metal thickness, and for other purposes. Radiation sources are also used by medical, in process operations (level gauges, on-stream analyzers, etc.), in the laboratory (X-ray diffraction unit), and by others. Where appropriate, a radiation control program should be in place and one individual should have overall coordination responsibility of the program. Licenses and records (wipe test results, monitoring results, etc.) should be properly maintained. A periodic review of radioactive material/ionizing radiation producing equipment use should be carried out with employees and contractors to ensure compliance with applicable regulations and accepted standards. Surveys should be conducted periodically to ensure exposure to ionizing radiation is as low as reasonably achievable.				
Comment					
Lighting	Lighting in operating areas, buildings, offsite locations, shops, and other areas must be adequate to enable personnel to work effectively without increased risk of injury. Day and nighttime lighting levels should be evaluated periodically to ensure established standards are met. An on-going lighting maintenance program should be in place.				
Comment					

* Based on Judgement: Y-Yes; N-No.

** Based on Judgement with Reference to the Information Presented under the Considerations Heading: Y-Yes; N-No.

*** Based on Judgement for Development/Implementation: H-High; M-Medium; L-Low.

ELEMENT	CONSIDERATIONS	PROGRAM ELEMENT			
		NEEDED* Y/N	IN PLACE* Y/N	ACCEPTABLE** Y/N	PRIORITY*** H/M/L
TECHNICAL ASPECTS (Continued)					
Monitoring Equipment	As appropriate, property maintained and calibrated personal and portable monitoring equipment should be available for assessing exposure of personnel to chemical/physical/biological agents and ergonomic stresses. Instruments should be available for collecting personal samples (e.g., for benzene, asbestos, lead, etc.); determining area/source contaminant concentrations (e.g., chlorine, hydrogen sulfide, carbon monoxide, etc.); and measuring noise levels, ventilation, illumination, ionizing radiation, etc. Instrument operating manuals should be accessible and records of instrument calibrations effectively documented.				
Comment					
Monitoring Program	A monitoring program and database should be in place for assessing exposures of personnel to potential health hazards. Repeat monitoring should be carried out, the extent of which will be dependent on professional judgement, exposure results, employee complaints/concerns, regulatory requirements, impending regulations, and other reasons. Qualitative reviews of job tasks, work practices and likely exposures to potentially hazardous materials, physical agents, biological agents, and ergonomic stresses should be conducted, as warranted, to identify exposure concerns. Where quantitative exposure measurements are necessary to assess risk, they should be conducted promptly and repeated, as appropriate, until the risks are controlled effectively and exposures are acceptable. Employees should be advised of quantitative monitoring results.				
Comment					

* Based on Judgement: Y-Yes; N-No.

** Based on Judgement with Reference to the Information Presented under the Considerations Heading: Y-Yes; N-No.

*** Based on Judgement for Development/Implementation: H-High; M-Medium; L-Low.

ELEMENT	CONSIDERATIONS	PROGRAM ELEMENT NEEDED* Y/N	IN PLACE* Y/N	ACCEPTABLE** Y/N	PRIORITY*** H/M/L
TECHNICAL ASPECTS (Continued)					
Non-Ionizing Radiation	Significant sources of non-ionizing radiation should be evaluated relative to their potential to present a health hazard to employees. Personnel should be provided appropriate training to make them aware of the potential hazards associated with non-ionizing radiation sources, such as ultraviolet and infrared radiation, microwaves, radar, etc. Where appropriate, engineering controls should be installed and personal protective equipment issued for minimizing exposure.				
Comment	______				
Sample Analysis	Samples obtained for assessing employee exposure to contaminants such as benzene, asbestos, lead, etc. should be collected by established procedures and submitted to a qualified laboratory (such as the TSL Industrial Hygiene Testing Laboratory) for analysis. Unidentified blank samples should be submitted routinely with the field samples and, where practical, replicates and spiked samples. The laboratory performing the analyses should be accredited and have an acceptable quality assurance program. A quality assurance program also should be in place for samples that are analyzed in-house (e.g., asbestos counting/identification, weight determinations, etc.).				
Comment	______				

* Based on Judgement: Y-Yes; N-No.

** Based on Judgement with Reference to the Information Presented under the Considerations Heading: Y-Yes; N-No.

*** Based on Judgement for Development/Implementation: H-High; M-Medium; L-Low.

ELEMENT	CONSIDERATIONS	PROGRAM ELEMENT			
		NEEDED* Y/N	IN PLACE* Y/N	ACCEPTABLE** Y/N	PRIORITY*** H/M/L
TECHNICAL ASPECTS (Continued)					
Ventilation	Local exhaust ventilation systems are used to minimize the release of materials from operations and thereby reduce employee exposure to airborne contaminants. Such systems are typically installed in mechanical shops, the laboratory, engine test (knock-lab) facility, utilities, and other locations. These systems should be designed properly and periodically tested to ensure design specifications are met. System designs should be reviewed relative to workplace health protection concerns before installation and an evaluation of system performance conducted at start-up. General ventilation system designs also should be reviewed before installation and performance evaluated periodically thereafter to ensure adequacy. System performance records should be maintained for reference.				
Comment					
SUPPORT ACTIVITIES					
Acquisitions	If a major acquisition is to be made, management should determine whether a detailed review of the operation should be conducted. Management should direct the detailed review to identify workplace health concerns; evaluate monitoring results; assess exposure potential; determine the status of compliance with applicable regulations; and evaluate other workplace health protection related issues.				
Comment					

* Based on Judgement: Y-Yes; N-No.

** Based on Judgement with Reference to the Information Presented under the Considerations Heading: Y-Yes; N-No.

*** Based on Judgement for Development/Implementation: H-High; M-Medium; L-Low.

ELEMENT	CONSIDERATIONS	NEEDED* Y/N	IN PLACE* Y/N	ACCEPTABLE** Y/N	PRIORITY*** H/M/L
SUPPORT ACTIVITIES (Continued)					
Contractor Support	Contractors need to be aware of the adverse health effect potential associated with their work in a facility, the measures to take to reduce exposure risk, and of the necessity to adhere to established procedures/work practices. Information (e.g., MSDSs, accident prevention manual, etc.) needs to be provided to contractors for this purpose. On some occasions, it may be practical for company personnel (contract coordination, industrial hygiene, safety) to carry out an oversight function and provide technical advice, as needed, to assure accepted/established procedures are adhered to by the contractor. The contractor should inform the facility of hazardous materials to be used in their work on the site, obtain a permit to use these, and demonstrate to the facility their compliance with applicable regulatory and facility requirements.				
Comment					
Follow-up of Recommendations	A procedure should be in place for follow-up of previously submitted workplace health protection recommendations to ensure that, when implemented, they are effective for reducing, or eliminating, an unacceptable exposure situation. Responsibility for carrying this out should be defined when the recommendation is made, along with reporting procedures for accountability purposes. Documentation of each follow-up assessment should be complete and readily retrievable.				
Comment					

* Based on Judgement: Y-Yes; N-No.
** Based on Judgement with Reference to the Information Presented under the Considerations Heading: Y-Yes; N-No.
*** Based on Judgement for Development/Implementation: H-High; M-Medium; L-Low.

ELEMENT	CONSIDERATIONS	PROGRAM ELEMENT NEEDED* Y/N	IN PLACE* Y/N	ACCEPTABLE** Y/N	PRIORITY*** H/M/L
SUPPORT ACTIVITIES (Continued)					
Hazard Awareness	Training and information programs should be established to make employees aware of the hazards associated with their work, the potential adverse health effects, control methods in place to minimize exposure, and measures that can be taken to reduce the potential for excessive exposure to hazardous substances/agents. Periodic retraining should be provided, as required, to maintain awareness. Training records should be maintained.				
Comment					
Investigations	Investigations should be conducted to address workplace health related issues which arise. Issues to be investigated may include employee complaints, accidental releases of hazardous materials, occupational disease occurrences, excessive exposure situations, spills, etc.				
Comment					

* Based on Judgement: Y-Yes; N-No.
** Based on Judgement with Reference to the Information Presented Under the Considerations Heading: Y-Yes; N-No.
*** Based on Judgement for Development/Implementation: H-High; M-Medium; L-Low.

ELEMENT	CONSIDERATIONS	NEEDED* Y/N	IN PLACE* Y/N	ACCEPTABLE** Y/N	PRIORITY*** H/M/L
SUPPORT ACTIVITIES (Continued)					
Permits	Workplace health protection issues should be addressed in the permit (hot work/cold work/safe work/confined space entry) issuance procedure. A determination should be made as to whether noise, radiation, or airborne contaminants are present at levels of health concern, and exposure to such hazards controlled, if necessary, before a permit is issued. Personnel making these measurements must know how to select and operate appropriate monitoring equipment, and to interpret the results.				
Comment					
Project Review	Plans for new facilities and modifications to existing ones should be reviewed in detail with respect to workplace health protection considerations. This is achieved most effectively by establishing a review and sign-off procedure for getting input from concerned groups, including the industrial hygiene function. The information developed should be provided to project management as early as possible in the design stage. Additional reviews may be warranted at various stages of construction. Start-up reviews are essential to ensure control measures are in place and effective.				
Comment					

(PROGRAM ELEMENT: NEEDED, IN PLACE, ACCEPTABLE, PRIORITY)

* Based on Judgement: Y-Yes; N-No.
** Based on Judgement with Reference to the Information Presented Under the Considerations Heading: Y-Yes; N-No.
*** Based on Judgement for Development/Implementation: H-High; M-Medium; L-Low.

ELEMENT	CONSIDERATIONS	PROGRAM ELEMENT			
		NEEDED* Y/N	IN PLACE* Y/N	ACCEPTABLE** Y/N	PRIORITY*** H/M/L
SUPPORT ACTIVITIES (Continued)					
Purchasing	A procedure should be in place to prevent the introduction of new materials into a facility that could result in the exposure of personnel to unknown hazards. The industrial hygiene function or others with workplace health protection program responsibility can assist by reviewing and approving purchase requisitions. Users of the material being purchased should be informed of special handling and exposure control requirements. This procedure will provide adequate lead time for implementing controls and providing employee hazard awareness training.				
Comment					
Recordkeeping	Monitoring results, survey findings, and results of investigations should be maintained for reference, statistical analysis, epidemiological studies, medical related issues, and for other purposes. Training program records also should be maintained. Those providing industrial hygiene support should have responsibility for maintaining monitoring results; reports of survey findings; investigation findings; ionizing radiation source information; and attendance records for those training programs it provides.				
Comment					

* Based on Judgement: Y-Yes; N-No.

** Based on Judgement with Reference to the Information Presented Under the Considerations Heading: Y-Yes; N-No.

*** Based on Judgement for Development/Implementation: H-High; M-Medium; L-Low.

ELEMENT	CONSIDERATIONS	PROGRAM ELEMENT: NEEDED* Y/N	IN PLACE* Y/N	ACCEPTABLE** Y/N	PRIORITY*** H/M/L
SUPPORT ACTIVITIES (Continued)					
Regulations	Responsibility should be assigned for maintaining an awareness of workplace health protection regulations; advising management on the impact of those regulations of an occupational health nature that are related to operations; developing a plan to respond to workplace health protection concerns; carrying out the plan; and advising management of the status of compliance.				
Comment	______________________				
PERSONAL PROTECTIVE EQUIPMENT					
Deluge Shower/ Eyewash Fountain	Deluge showers/eyewash fountains, when required, should be located appropriately within the emergency use. Personnel should be trained in their use. Equipment must be readily accessible, meet relevant design specifications, be adequately identified, properly maintained, and periodically inspected and tested.				
Comment	______________________				

* Based on Judgement: Y-Yes; N-No.
** Based on Judgement with Reference to the Information Presented under the Considerations Heading: Y-Yes; N-No.
*** Based on Judgement for Development/Implementation: H-High; M-Medium; L-Low.

ELEMENT	CONSIDERATIONS	PROGRAM ELEMENT			
		NEEDED* Y/N	IN PLACE* Y/N	ACCEPTABLE** Y/N	PRIORITY*** H/M/L
PERSONAL PROTECTIVE EQUIPMENT (Continued)					
Hearing Protection	Personal hearing protective equipment must be adequate for effectively reducing the noise exposure of personnel below levels of health concern. Personnel must be trained in its use, and these devices should be readily available for use in previously identified high noise areas.				
Comment					
Other Protective Equipment	Protective equipment for preventing excessive exposure to hazardous materials/agents must be properly selected and maintained and be readily available to personnel who have been trained in its use.				
Comment					

* Based on Judgement: Y-Yes; N-No.

** Based on Judgement with Reference to the Information Presented under the Considerations Heading: Y-Yes; N-No.

*** Based on Judgement for Development/Implementation: H-High; M-Medium; L-Low.

ELEMENT	CONSIDERATIONS	PROGRAM ELEMENT NEEDED* Y/N	IN PLACE* Y/N	ACCEPTABLE** Y/N	PRIORITY*** H/M/L
PERSONAL PROTECTIVE EQUIPMENT (Continued)					
Respiratory Protection	If respiratory protection is needed and/or used by personnel, a written respiratory protective equipment program should be established. Employees should be fit-tested and medically evaluated prior to their use of this equipment. The function assigned industrial hygiene responsibility should provide technical support in the development, implementation, and maintenance of the respirator program. A program administrator should be identified, and the written program should address the selection of equipment, inspection, maintenance, availability, cleaning, storage, and training of personnel in its proper use. Records of training should be maintained. The program should be evaluated periodically (annually is suggested) to ensure effectiveness.				
Comment					
SPECIAL MATERIALS					
Alkyl Lead Compounds	Adherence with accepted procedures, work practices and personal protective equipment use is necessary to prevent excessive exposure to alkyl lead compounds. In addition, the posting of leaded fuel storage tanks, the handling and disposal of leaded sludge, cleaning of leaded gasoline storage tanks, and the disposal of lead and/or reuse of contaminated metals are issues of concern. Personnel potentially exposed to alkyl lead compounds should be included in a biological test program.				
Comment					

* Based on Judgement: Y-Yes; N-No.

** Based on Judgement with Reference to the Information Presented under the Considerations Heading: Y-Yes; N-No.

*** Based on Judgement for Development/Implementation: H-High; M-Medium; L-Low.

ELEMENT	CONSIDERATIONS	NEEDED* Y/N	IN PLACE* Y/N	ACCEPTABLE** Y/N	PRIORITY*** H/M/L
SPECIAL MATERIALS (Continued)					
Aromatic Process Oils	Exposure of personnel to aromatic process oils is a workplace health concern in a number of petroleum and petrochemical operations, such as in cracking units (catalytic, steam, thermal, etc.) and lube extraction. Effective procedures and work practices, as well as the use of personal protection, are necessary to prevent skin contact with these oils or the inhalation of vapors/mists.				
Comment					
Asbestos	If asbestos-containing material (ACM) is present on site, a program should be in place which identifies locations of use; condition of material; exposure monitoring needs; procedures/work practices for handling/removal/disposal; regulatory compliance requirements; respirator program needs; review of contract bid specifications; contractor oversight; etc.				
Comment					

* Based on Judgement: Y-Yes; N-No.
** Based on Judgement with Reference to the Information Presented under the Considerations Heading: Y-Yes; N-No.
*** Based on Judgement for Development/Implementation: H-High; M-Medium; L-Low.

ELEMENT	CONSIDERATIONS	NEEDED* Y/N	IN PLACE* Y/N	ACCEPTABLE** Y/N	PRIORITY*** H/M/L
SPECIAL MATERIALS (Continued)					
Compressed Gases	Compressed gases used in petroleum and petrochemical operations may be flammable and/or toxic. The potential for exposure to these chemicals at toxic levels (parts per million by volume) must be evaluated. Handling, storage, and use concerns related to the toxic properties of compressed gases need to be addressed. Storage area ventilation must be adequate. The possible need for installation of continuous monitors should be evaluated for materials of high toxicity.				
Comment					
Sour Streams	Special procedures must be established and followed where there is a potential for exposure to H_2S containing streams/products. This involves hazard awareness training; special sampling/gauging procedures; purging and venting of equipment containing such materials; and the use of appropriate respiratory protection, the buddy system, and emergency response considerations. Hazard awareness, emergency, and work practice training are necessary for all personnel potentially exposed to H_2S.				
Comment					

* Based on Judgement: Y-Yes; N-No.

** Based on Judgement with Reference to the Information Presented under the Considerations Heading: Y-Yes; N-No.

*** Based on Judgement for Development/Implementation: H-High; M-Medium; L-Low.

ELEMENT	CONSIDERATIONS	PROGRAM ELEMENT NEEDED* Y/N	IN PLACE* Y/N	ACCEPTABLE** Y/N	PRIORITY*** H/M/L
SPECIAL OPERATIONS					
Abrasive Cleaning	Where abrasive cleaning operations are carried out, a program should be in place that addresses the type of abrasive(s) used; potential for exposure to lead, silica, etc.; adequacy of personal protective equipment being used; quality of breathing air; effectiveness of respiratory protection; ventilation system assessment and maintenance; medical surveillance; etc. If contracted abrasive cleaning is done, the facility should identify potential exposure concerns to the contractor (e.g., leaded paint, vapors/gas exposure potential, emergency alarms, evacuation procedures, work permit practices, etc.).				
Comment					
Bulk Product Loading	Personnel may be exposed to vapor/mist emissions during open loading of liquid petroleum and petrochemical products. Established procedures/work practices should be adhered to by personnel engaged in this work, and exposure controls implemented, where warranted, based on an evaluation of the potential hazards and/or exposure monitoring results.				
Comment					

* Based on Judgement: Y-Yes; N-No.

** Based on Judgement with Reference to the Information Presented under the Considerations Heading: Y-Yes; N-No.

*** Based on Judgement for Development/Implementation: H-High; M-Medium; L-Low.

ELEMENT	CONSIDERATIONS	NEEDED* Y/N	IN PLACE* Y/N	ACCEPTABLE** Y/N	PRIORITY*** H/M/L
SPECIAL OPERATIONS (Continued)					
Confined Space Entry	Entry into a confined space should require an entry permit. Work in such a location may require isolation and ventilation of the space; carrying out appropriate tests (in the proper manner for the contaminants of concern); requiring the use of appropriate protective equipment; providing training for personnel entering the space, as well as for the standby person; and ensuring that all confined space work requirements are followed.				
Comment					
Fugitive Emissions	Exposure to airborne contaminants in petroleum and petrochemical operations often is the result of releases from a multitude of sources. If not effectively controlled, such sources can contribute significantly to personnel exposures. These should be identified during surveys/investigations, and measurements carried out to determine whether releases from pumps, valves, regulators, sewer vents, etc. are potentially harmful or need to be reduced.				
Comment					

(PROGRAM ELEMENT spans NEEDED, IN PLACE, ACCEPTABLE, PRIORITY.)

* Based on Judgement: Y-Yes; N-No.
** Based on Judgement with Reference to the Information Presented under the Considerations Heading: Y-Yes; N-No.
*** Based on Judgement for Development/Implementation: H-High; M-Medium; L-Low.

ELEMENT	CONSIDERATIONS	NEEDED* Y/N	IN PLACE* Y/N	ACCEPTABLE** Y/N	PRIORITY*** H/M/L
SPECIAL OPERATIONS (Continued)					
Hazard Warning Signs	Warning signs should be posted to alert personnel of the presence of hazards in a process or at various locations/areas. The signs should be placed so they are effective for alerting personnel of the potential hazard, are readily visible, provide a clear message, and are uniform (color, wording, etc.) throughout the facility. Damaged or defaced signs should be replaced.				
Comment	____________				
Laboratory Operations	Many hazardous chemicals are used in laboratory procedures. Operations should be reviewed to ensure handling procedures, engineering controls, work practices, protective equipment, labeling practices, hazard awareness, and other measures are effective for minimizing employee exposure to the materials handled.				
Comment	____________				

(PROGRAM ELEMENT spans the NEEDED, IN PLACE, ACCEPTABLE and PRIORITY columns.)

* Based on Judgement: Y-Yes; N-No.
** Based on Judgement with Reference to the Information Presented under the Considerations Heading: Y-Yes; N-No.
*** Based on Judgement for Development/Implementation: H-High; M-Medium; L-Low.

ELEMENT	CONSIDERATIONS	NEEDED* Y/N	IN PLACE* Y/N	ACCEPTABLE** Y/N	PRIORITY*** H/M/L
		PROGRAM ELEMENT			
SPECIAL OPERATIONS (Continued)					
Mechanical Shops	Shop operations such as welding, pump maintenance, valve adjustment/testing, metallizing, lapping, gasket cutting, lead burning, machining, and others can result in the generation of toxic substances, physical agents, or bioorganisms which can cause adverse health effects among exposed personnel. Exposure controls must be effective for preventing such occurrences. Periodic reviews of shop operations should be conducted to ensure that controls are effective and excessive exposures do not occur.				
Comment	______				
Packaging	Personnel may be exposed to vapors and mists emitted during product blending, packaging, and warehousing. Inhalation and/or skin contact hazards may exist. Noise levels may be elevated (greater than 85 dBA). The carbon monoxide concentration may be excessive where internal combustion powered equipment is employed to transfer product or for housekeeping. Exposure controls should be implemented, where warranted.				
Comment	______				

* Based on Judgement: Y-Yes; N-No.
** Based on Judgement with Reference to the Information Presented under the Considerations Heading: Y-Yes; N-No.
*** Based on Judgement for Development/Implementation: H-High; M-Medium; L-Low.

ELEMENT	CONSIDERATIONS	NEEDED* Y/N	IN PLACE* Y/N	ACCEPTABLE** Y/N	PRIORITY*** H/M/L
		PROGRAM ELEMENT			
SPECIAL OPERATIONS (Continued)					
Process and Product Sampling Operations	The potential for exposure can be significant during process or product sampling activities. Sample line flushing and sample vessel purging can result in the spillage/venting of raw material/intermediate/product with consequent exposure to the contaminant via inhalation and/or skin contact. Proper sampling practices should be followed, employee exposures evaluated and, where necessary, controls implemented to reduce exposure potential. Closed loop sampling could be employed, where appropriate.				
Comment					
Receipt and Storage of Materials	A procedure should be in place to assure that purchase requisitions are reviewed to determine whether materials to be procured can be handled/used without excessive risk to personnel. In addition, Material Safety Data Sheets should be obtained from the supplier and made available to personnel who have been informed of the potential health hazards associated with exposure to the substance. Materials should be stored so as not to present a potential incompatibility with other materials in storage/use.				
Comment					

* Based on Judgement: Y-Yes; N-No.

** Based on Judgement with Reference to the Information Presented under the Considerations Heading: Y-Yes; N-No.

*** Based on Judgement for Development/Implementation: H-High; M-Medium; L-Low.

ELEMENT	CONSIDERATIONS	PROGRAM ELEMENT NEEDED* Y/N	IN PLACE* Y/N	ACCEPTABLE** Y/N	PRIORITY*** H/M/L
SPECIAL OPERATIONS (Continued)					
Tank Cleaning	Tank cleaning practices must be well established and adhered to with regard to opening, working from the outside, vacuuming-up residue/sludge, ventilating, and testing for the presence of toxic contaminants before tank entry. Additional testing (O_2, hydrocarbons, benzene, H_2S, etc.) may have to be conducted while work is proceeding. Concentrations of contaminants must be determined before entry is permitted. Based on contaminant concentration, personnel may have to wear respiratory protection. They should be properly trained with regard to working in a confined space and in the use of the respiratory protection.				
Comment					
Turnarounds	The potential for exposure to chemical, physical and biological agents can be greater during turnaround (T/A) activities of process operating units than during routine operations. Thus, there may be a need to review procedures and work practices, monitor personnel and document exposures during T/A activities. Personnel with industrial hygiene responsibility should be aware of the T/A schedule and the scope of the work to be done, so that monitoring plans can be developed and submitted to facility management for approval. The results of previous T/A exposure monitoring sessions should be reviewed and the findings used as a basis for developing monitoring plans and exposure control strategies for future T/A work.				
Comment					

* Based on Judgement: Y-Yes; N-No.
** Based on Judgement with Reference to the Information Presented under the Considerations Heading: Y-Yes; N-No.
*** Based on Judgement for Development/Implementation: H-High; M-Medium; L-Low.

ELEMENT	CONSIDERATIONS	PROGRAM ELEMENT: NEEDED* Y/N	IN PLACE* Y/N	ACCEPTABLE** Y/N	PRIORITY*** H/M/L
SPECIAL OPERATIONS (Continued)					
Visual Display Terminals	Visual display terminals which are widely used in the petroleum and petrochemical industries may have the potential to cause adverse health effects. Such effects may be the result of work-station design, inadequate lighting, glare, trauma due to repetitive motion, etc. The potential for such effects to occur should be evaluated and controlled, as necessary.				
Comment					
EMERGENCY PREPAREDNESS					
Emergency Response	Response plans for emergency situations should be reviewed to ensure workplace health considerations have been incorporated. Plan development, exposure monitoring considerations, personal protective equipment selection, toxicity considerations, training program development, and plan implementation may need industrial hygiene input. The emergency response plans should be reviewed, as necessary, to ensure they have addressed newly introduced hazards and are responsive to associated needs. Where established, the industrial hygiene function should be an active participant in the periodic assessment of this program element.				
Comment					

* Based on Judgement: Y-Yes; N-No.

** Based on Judgement with Reference to the Information Presented under the Considerations Heading: Y-Yes; N-No.

*** Based on Judgement for Development/Implementation: H-High; M-Medium; L-Low.

ELEMENT	CONSIDERATIONS	NEEDED* Y/N	IN PLACE* Y/N	ACCEPTABLE** Y/N	PRIORITY*** H/M/L
MEDICAL SURVEILLANCE					
Audiometric Test of Results	The effectiveness of a hearing conservation program at a facility is determined from the results the audiometric testing of employees. Permanent hearing threshold shifts among personnel may indicate a deficiency in the noise exposure control program. On-going procedures should be in place to promptly observe any occurrence of occupational hearing loss. If observed, it must trigger effective communication among medical, facility management, and personnel with industrial hygiene responsibility, so that measures can be implemented promptly to prevent further hearing loss.				
Comment					
Biological Test Results	Biological testing may be performed to determine if employees have been excessively exposed to hazardous substances such as benzene, lead, ionizing radiation or others. If an indication of an excessive exposure is detected, an investigation should be initiated to identify the likely cause. Measures should be implemented promptly to prevent a recurrence. Procedures should be in place for prompt follow-up to ensure that the hazard has been reduced.				
Comment					

PROGRAM ELEMENT

* Based on Judgement: Y-Yes; N-No.
** Based on Judgement with Reference to the Information Presented Under the Considerations Heading: Y-Yes; N-No.
*** Based on Judgement for Development/Implementation: H-High; M-Medium; L-Low.

4F:WS1211a.doc

2. MANAGEMENT

WORKPLACE HEALTH PROTECTION PROGRAM ELEMENT
ON
ADMINISTRATION

WORKPLACE HEALTH PROTECTION PROGRAM ELEMENT ON ADMINISTRATION

OBJECTIVE

To provide information regarding a methodology for identifying workplace health protection program element needs and a strategy for their development and implementation.

BACKGROUND

Workplace health protection program administration involves the planning and directing of the activities of those functions responsible for the development, implementation, and maintenance of those elements management considers necessary for achieving workplace health protection goals. Functional plans that are developed to achieve these goals should be reviewed and approved by management, and the activities of the responsible groups periodically stewarded to management.

CONSIDERATIONS

A written policy statement should exist indicating management's intention to ensure that the health hazards present in the workplace will be controlled such that adverse health effects will not occur among personnel engaged in work activities within the facility or to potentially exposed residents in the surrounding community. The policy should be communicated to all employees and responsibilities with regard to stated goals defined. Where available, the industrial hygiene function, in cooperation with other groups (e.g., medical, safety, environmental, etc.), should, with management support, ensure that the policy is implemented.

An initial step in the administration of a workplace health protection program is to develop information which management can use to make judgments regarding the applicability of, and need for developing and implementing specific program elements. An assessment of the existing workplace health protection program should be carried out to identify such needs. The table in the Introduction can be employed to develop information for management consideration. If more detail is necessary, the specific guidance checklists that are part of each of the program elements presented here can be completed and a summary of the findings provided to management.

When necessary, the program elements identified by management to be a part of the facility's workplace health protection program could then be developed and implemented. If deficiencies are noted in existing program elements, it will be necessary to identify needs, and to implement recommendations for corrective action to make them more effective. Both will require the planning and directing of the activities of available resources to accomplish the goals defined in the policy statement.

A strategy will be necessary for program element development and implementation to accomplish goals. The issues addressed in the considerations section of each of the elements can form the basis for program planning and development. Knowledge of the operations, as well as the jobs and tasks performed by personnel, is essential for effective program development. Information obtained in a qualitative exposure assessment effort can be of benefit in understanding these factors. In addition, communication with the various groups to be affected is essential during program development to obtain agreement as to specific needs, method of implementation, and identifying support/resources for ensuring that the program will be effectively maintained.

Periodic reviews (audits) of the workplace health protection program elements are advisable to determine their effectiveness and to identify needs. A program review frequency of every 3 years by an outside group would be warranted. In addition, reviews by in-house personnel every 2 years is suggested. The checklist presented herein for each of the program elements can be employed by both groups in these reviews. If appropriate, each checklist could be modified to address specific regulatory requirements. The reviews should be considered constructive and the information developed used to achieve the overall objectives of the workplace health protection program. The findings of these reviews will be of benefit to management, the industrial hygiene function, and personnel whose health and wellness are being protected by the workplace health protection program.

REFERENCES

Garrett, J. T., et. al., "Industrial Hygiene Management (New York: John Wiley & Sons, Inc., 1988)

Blakeslee, H. W. and T. M. Grabowski, "A Practical Guide to Plant Environmental Audits" (New York: Van Nostrand Reinhold Company, 1985).

Largent, E. J. and J. B. Olishifski, "Fundamentals of Industrial Hygiene", 2nd Edition (Chicago, IL: National Safety Council, 1985).

"Self Evaluation of Occupational Safety and Health Programs" (Cincinnati, OH: U.S. Department of Health, Education, and Welfare, National Institute for Occupational Safety and Health, 1978).

WORKPLACE HEALTH PROTECTION PROGRAM ELEMENT CHECKLIST ON ADMINISTRATION

		Yes	No	Don't Know	N/A
1.	Is there a policy statement regarding workplace health protection program considerations?	___	___	___	___
2.	Is the policy regarding workplace health protection communicated to all employees?	___	___	___	___
3.	Are responsibilities related to work-place health protection program goals clearly defined, particularly for functions that have accountability for specific program elements?	___	___	___	___
4.	Is the role of the industrial hygiene function defined in writing?	___	___	___	___
5.	Does the industrial hygiene function have management support to develop an effective comprehensive work-place health protection program?	___	___	___	___
6.	Is an annual industrial hygiene function budget prepared for management review and approval?	___	___	___	___
7.	Are the plans and activities of the industrial hygiene function:				
	. Reviewed by management?	___	___	___	___
	. Approved by management?	___	___	___	___
	. Periodically stewarded to management?	___	___	___	___
8.	Has a review been conducted to evaluate the effectiveness of the workplace health protection program?	___	___	___	___
9.	Based on review findings, were all applicable workplace health protection program elements in place and effective?	___	___	___	___

N/A - Not Applicable

WORKPLACE HEALTH PROTECTION PROGRAM ELEMENT CHECKLIST ON ADMINISTRATION

		Yes	No	Don't Know	N/A
10.	Are periodic reviews of the workplace health protection program carried out by:				
	. Management?	___	___	___	___
	. In-house group?	___	___	___	___
	. External group?	___	___	___	___
	. Consultant?	___	___	___	___
11.	Has the information developed in workplace health protection program reviews been communicated to management?	___	___	___	___
12.	Has an effort been made to make management aware of all elements which make up a comprehensive workplace health protection program?	___	___	___	___
13.	Has management communicated with responsible functional groups regarding those program elements that are to be developed, implemented or improved?	___	___	___	___
14.	Is the workplace health protection program considered effective by:				
	. Industrial hygiene?	___	___	___	___
	. Management?	___	___	___	___
	. Medical?	___	___	___	___
	. Safety?	___	___	___	___
15.	Are there sufficient resources available to develop, implement, and maintain an effective, comprehensive workplace health protection program?	___	___	___	___
16.	Are program element needs discussed with affected groups during development?	___	___	___	___
17.	Are personnel who are responsible for each element of the workplace health protection program adequately trained to develop, implement and maintain it?	___	___	___	___

N/A - Not Applicable

WORKPLACE HEALTH PROTECTION PROGRAM ELEMENT CHECKLIST ON ADMINISTRATION

	Yes	No	Don't Know	N/A
18. Is there an effort to obtain additional resources when needed?	___	___	___	___
19. Have management personnel attended the facility's hazard awareness training program?	___	___	___	___
20. Does management have an awareness of all health hazards associated with operations in the facility?	___	___	___	___
21. Is there direct communication between the industrial hygiene function and top management?	___	___	___	___
22. Has the industrial hygiene group leader been provided management training by the company?	___	___	___	___
23. Have industrial hygiene personnel been provided training in:				
. Basic industrial hygiene?	___	___	___	___
. Ergonomics?	___	___	___	___
. Hearing conservation?	___	___	___	___
. Heat stress?	___	___	___	___
. Industrial ventilation?	___	___	___	___
. Ionizing radiation?	___	___	___	___
. Noise measurement and control?	___	___	___	___
. Non-ionizing radiation?	___	___	___	___
. Process technology (re: operations being carried out)?	___	___	___	___
. Respiratory protection?	___	___	___	___
. Sampling and analysis?	___	___	___	___
. Other? (______________________)	___	___	___	___
24. Are drafts of industrial hygiene reports reviewed by the addressee prior to issue?	___	___	___	___
25. Are reports routinely reviewed for technical accuracy before issue?	___	___	___	___

N/A - Not Applicable

WORKPLACE HEALTH PROTECTION PROGRAM ELEMENT CHECKLIST ON ADMINISTRATION

	Yes	No	Don't Know	N/A
26. Is the performance of the industrial hygiene staff evaluated periodically?	___	___	___	___

N/A - Not Applicable

WORKPLACE HEALTH PROTECTION PROGRAM ELEMENT
ON
COMMUNICATIONS

WORKPLACE HEALTH PROTECTION PROGRAM ELEMENT ON COMMUNICATIONS

OBJECTIVE

To identify communication considerations which facilitate the development, implementation, and maintenance of a comprehensive workplace health protection program.

BACKGROUND

Communications is an essential part of health protection programs. For example, the media attention given radiological health issues in the 1940s and 1950s and that given environmental health in the 1970s and 1980s was effective in alerting individuals, communities, government agencies, and industry of the potential adverse health effects and environmental damage that could result from ineffective control of hazardous materials or agents. As a result, there was a public reaction, on an international scale, for the implementation of measures to minimize the potential for adverse effects to occur.

A lesson to be learned from those experiences is that an effective comprehensive workplace health protection program will not evolve without identifying the need, and communicating this to all involved persons. Thus, communications must be established and maintained with all levels of an organization to obtain the support necessary to develop, implement, and maintain a comprehensive workplace health protection program.

Communications must be established and maintained between the industrial hygiene function, management, supervisors, environmental personnel, design groups, committees, emergency forces, contractors, and others. Each must be aware that those groups with functional responsibility for various elements of the workplace health protection program can assist in providing support necessary to implement and maintain an effective program that is essential for preventing adverse occupationally related health effects among personnel.

CONSIDERATIONS

There should be a policy statement, endorsed and supported by management, regarding workplace health protection. Its contents should be communicated to all employees.

Workplace health protection program plans should be based on input from the various functions of the organization. The plan should be communicated to management for review and approval. Activities carried out to meet plan objectives should be periodically stewarded to management. In addition, management should be kept apprised of problems in meeting goals, of reordered priorities, the potential impact of impending regulations, resource needs, etc.

Groups with workplace health protection responsibility should establish and maintain communications with the operations management group. Supervisors must be aware of the potential health hazards associated with materials to which personnel may be exposed in their areas of responsibility, as well as of the health hazards associated with the activities which personnel under their supervision perform. Industrial hygiene should respond promptly to their requests for support related to these issues.

Industrial hygiene and safety personnel should routinely interact on issues that impact both functions. In addition, industrial hygiene personnel should serve on committees with safety department personnel. Communications between these two groups must be ongoing for effective resolution of common concerns.

There are a number of opportunities for personnel with workplace health protection program responsibility to establish communications with employees. These occur during hazard communication training, respiratory protective equipment training, when carrying out personal monitoring programs, during periodic surveys and program reviews, as well as during inspections/investigations.

Communication between medical department personnel and the industrial hygiene function is essential for the development, implementation, and maintenance of the hearing conservation program, the occupational health investigation effort, the biological monitoring program, substance specific medical surveillance programs, as well as other aspects of the overall workplace health protection effort. Since the aforementioned program elements generally involve human resource considerations, communication between industrial hygiene and this function is also necessary.

The control measures recommended by the industrial hygiene function impact facility support and engineering functions. Communications with these groups is necessary regarding equipment specifications, as well as recommended control measures and their implementation.

When investigations, surveys, program reviews or other industrial hygiene activities are to be carried out in specific areas of a facility, management and employees of the area should be made aware of the purpose of the activity and encouraged to participate, as appropriate. Monitoring data obtained in these activities should be discussed with management and the supervisor before the results are given to the employee by the supervisor.

Information exchange is necessary between those with workplace health protection program responsibility and various groups, such as the laboratory, contract coordination, inspection, purchasing, product movement and others. This will facilitate visits to their work areas to ensure appropriate procedures/work practices are being followed, proper protective equipment is being used, and to provide other advice related to workplace health protection issues, as warranted.

REFERENCES

Konikow, R. B. and F. E. McElroy, "Communications for the Safety Professional", (Chicago, IL: National Safety Council, 1975).

Lowry, G. C. and R. C. Lowry, "Handbook of Hazard Communications and OSHA Requirements", (Chelsea, MI: Lewis Publishers, Inc., 1985).

WORKPLACE HEALTH PROTECTION PROGRAM ELEMENT CHECKLIST ON COMMUNICATIONS

	Yes	No	Don't Know	N/A
1. Is there a written policy statement that addresses the workplace health protection program?	___	___	___	___
2. Have employees been informed of this policy?	___	___	___	___
3. Are industrial hygiene plans based on management's direction with regard to the scope of the workplace health protection program?	___	___	___	___
4. Is there communication and concurrence with various groups who will be provided support (through planned activities) before the plan is submitted to management for review and approval?	___	___	___	___
5. Does management:				
. Review industrial hygiene plans?	___	___	___	___
. Approve industrial hygiene plans?	___	___	___	___
. Require stewardship of activities against approved plans?	___	___	___	___
6. Is there an ongoing effort to keep management informed of:				
. Problems encountered in meeting goals?	___	___	___	___
. Changes in priorities?	___	___	___	___
. Problems in resolving issues?	___	___	___	___
. Impact of impending regulations?	___	___	___	___
. Resource needs to accomplish planned and unplanned activities?	___	___	___	___
7. Do industrial hygiene personnel know all the supervisors in the facility?	___	___	___	___
8. Do all supervisory personnel know the industrial hygienist?	___	___	___	___
9. Do industrial hygiene personnel respond promptly to supervisor:				
. Concerns?	___	___	___	___
. Requests for support?	___	___	___	___

N/A - Not Applicable

WORKPLACE HEALTH PROTECTION PROGRAM ELEMENT CHECKLIST ON COMMUNICATIONS

		Yes	No	Don't Know	N/A
10.	Are monitoring results discussed with management?	___	___	___	___
11.	Are monitoring results provided to management to give to their personnel?	___	___	___	___
12.	Is there effective communication between safety and industrial hygiene personnel in:				
	. Committee activities?	___	___	___	___
	. Employee training?	___	___	___	___
	. Investigations?	___	___	___	___
	. Problem solving?	___	___	___	___
13.	Is industrial hygiene represented on safety committees:				
	. Facility?	___	___	___	___
	. Interplant?	___	___	___	___
	. Laboratory?	___	___	___	___
	. Mechanical?	___	___	___	___
	. Process?	___	___	___	___
	. Other? (__________)	___	___	___	___
14.	Are efforts made by industrial hygiene to establish and maintain communications with operations, maintenance, inspection, and other facility personnel?	___	___	___	___
15.	Is the communication link with the medical department effective with regard to the:				
	. Asbestos medical surveillance program?	___	___	___	___
	. Benzene medical surveillance program?	___	___	___	___
	. Hearing conservation program?	___	___	___	___
	. Laboratory worker program?	___	___	___	___
	. Occupational health investigations?	___	___	___	___
	. Respirator use program?	___	___	___	___
	. Other? (______________)	___	___	___	___
16.	Is there effective interaction and communication between those with industrial hygiene responsibility and human resource personnel on matters of mutual interest?	___	___	___	___
17.	Is a mechanism in place to inform the industrial hygiene function when process equipment is to be opened?	___	___	___	___

N/A - Not Applicable

WORKPLACE HEALTH PROTECTION PROGRAM ELEMENT CHECKLIST ON COMMUNICATIONS

	Yes	No	Don't Know	N/A
18. Is there effective interaction between those with responsibility for workplace health protection program elements and potentially affected groups/individuals on measures to be implemented to reduce employee exposure to health hazards:				
. Management?	___	___	___	___
. Supervision?	___	___	___	___
. Facilities?	___	___	___	___
. Engineering?	___	___	___	___
. Other? (________________)	___	___	___	___
19. Are those with industrial hygiene responsibility kept informed of work to be done on site by contractors?	___	___	___	___
20. Does the industrial hygiene function participate in meetings with contractors and/or others:				
. Before a contract is let?	___	___	___	___
. To review contractor bids relative to workplace health issues?	___	___	___	___
. During the contracted work period?	___	___	___	___
. Post completion discussions on contractor performance?	___	___	___	___
21. Is the industrial hygiene function informed when:				
. Radiography is to be performed?	___	___	___	___
. Tank cleaning is to be done?	___	___	___	___
. Sludge is to be removed from wastewater facilities?	___	___	___	___
. Turnaround work is planned?	___	___	___	___
. A potentially hazardous task is to be done?	___	___	___	___
22. Do industrial hygiene and medical department personnel participate in joint investigations based on:				
. Biological monitoring results?	___	___	___	___
. Toxicology information?	___	___	___	___

N/A - Not Applicable

WORKPLACE HEALTH PROTECTION PROGRAM ELEMENT CHECKLIST ON COMMUNICATIONS

	Yes	No	Don't Know	N/A
. Exposure monitoring results?	___	___	___	___
. Medical surveillance findings?	___	___	___	___
23. Does purchasing contact industrial hygiene when a new chemical/material is to be purchased for use on site?	___	___	___	___
24. Are contractors required to inform industrial hygiene of the chemicals/products they will bring on site and provide a copy of the MSDS for these?	___	___	___	___
25. Is industrial hygiene required to review all MSDSs to ensure their adequacy?	___	___	___	___
26. Does management discuss the findings of internal/external workplace health protection program reviews (audits) with industrial hygiene personnel?	___	___	___	___
27. Do industrial hygiene personnel participate in presentations to new employees regarding workplace health protection issues?	___	___	___	___
28. Is there effective communication between the laboratory (i.e., that analyzes samples) and the industrial hygiene function?	___	___	___	___
29. Do field industrial hygiene personnel communicate with laboratory supervision, as needed, to:				
. Determine sample volumes necessary to obtain appropriate limits of detection?	___	___	___	___
. Determine sampling methods appropriate for a contaminant?	___	___	___	___
. Determine sample storage needs?	___	___	___	___
. Determine preservatives for samples?	___	___	___	___
. Request spiked samples?	___	___	___	___

N/A - Not Applicable

WORKPLACE HEALTH PROTECTION PROGRAM ELEMENT
ON
PLANNING/STEWARDSHIP

WORKPLACE HEALTH PROTECTION PROGRAM ELEMENT
ON
PLANNING/STEWARDSHIP

OBJECTIVE

To identify the need for the development, management review, and approval of a workplace health protection program, and periodic stewardship of activities to management on plan accomplishments/failures.

BACKGROUND

For many years the support provided in industry relative to workplace health protection issues was on an ad hoc basis. Generally, more time was spent in "fire-fighting" than in developing a proactive plan and carrying out the activities identified in it. In addition, when plans were developed, they, very likely, were not submitted to management for review and approval.

The impact of regulations on those responsible for various aspects of the workplace health protection program has been such that all available resources may be required to address regulatory requirements. Adequate resources may not be available to respond to these, as well as the day-to-day concerns which arise, and also develop, implement, and maintain other elements of the facility's workplace health protection program. Thus, a need has evolved requiring the effective management of those responsible for the various elements of the workplace health protection program, as well as the development, implementation, and maintenance of the overall program. This can be accomplished most effectively by defining program needs, developing an action plan, submitting it to management for review and approval, identifying groups responsible for each element's effectiveness and periodically steward activities to management. Management can then assess program acccomplishments, resource needs, and the measures necessary to ensure that the workplace health protection program is effective.

CONSIDERATIONS

A formal procedure can be implemented for the development of an annual plan which defines the goals and objectives of the workplace health protection program, as well as the planned activities of those groups responsible for its various elements. The plan should be submitted to management for review and approval. Each group/function impacted by the proposed activities identified in the plan must have the opportunity to review it and provide comment, as necessary, before it is submitted to management for approval.

The plan should be sufficiently flexible to enable responsible groups to respond to regulatory requirements, as well as the day-to-day issues that arise. Since the day-to-day requests and regulatory requirements can affect a group's ability to accomplish planned objectives, management needs to be aware of these demands so that additional resources can be provided, as warranted, to accomplish objectives.

The impact of new regulations and company program directives should be anticipated, if at all possible, and sufficient time incorporated into the plan to respond to the resulting demands on resources. Turnaround activities and/or construction work, which may impact the plan and resources, should be anticipated and additional personnel and equipment arranged for, if necessary, to respond to these future needs. Management should be aware of those situations which may impose excessive demands on available resources.

Information determined in qualitative exposure assessments will be useful in the identification of workplace health protection program needs. For example, shortcomings in hazard awareness training, engineering controls, personal protective equipment use, the monitoring effort, and others will be identified in a qualitative exposure assessment. Activities which address the identified areas/issues of concern can then be incorporated into the written workplace health protection plan for management review and approval and subsequent development and implementation.

REFERENCES

Garrett, J. T., et al, "Industrial Hygiene Management" (New York: John Wiley & Sons, 1988).

**WORKPLACE HEALTH PROTECTION PROGRAM ELEMENT CHECKLIST
ON
PLANNING/STEWARDSHIP**

		Yes	No	Don't Know	N/A
1.	Are workplace health protection program activities carried out on a planned basis?	___	___	___	___
2.	Is a written plan developed annually which addresses workplace health protection program activities?	___	___	___	___
3.	Are present activities designed to meet the objectives identified in a written plan?	___	___	___	___
4.	Is each year's plan reviewed and approved by management?	___	___	___	___
5.	Is there sufficient time allocated for plan development, review, and approval so that the plan is in place at the beginning of the planning period?	___	___	___	___
6.	Are the details of the plan communicated to those managements responsible for the areas/operations to be impacted by the planned activities?	___	___	___	___
7.	Does facility management communicate with supervisory personnel so that effective cooperation is provided in the conduct of planned activities?	___	___	___	___
8.	Are future turnaround plans communicated to the various functions with workplace health protection program responsibilities so that support needs can be anticipated?	___	___	___	___

N/A - Not Applicable

WORKPLACE HEALTH PROTECTION PROGRAM ELEMENT CHECKLIST ON PLANNING/STEWARDSHIP

		Yes	No	Don't Know	N/A
9.	Does the plan approval procedure serve as a means for obtaining support to meet objectives?	___	___	___	___
10.	Are activities periodically stewarded to management?	___	___	___	___
11.	Is a summary report on goals/ accomplishments/failures periodically provided to management?	___	___	___	___
12.	Is it practical and easy to modify the plan to respond to changing priorities?	___	___	___	___
13.	Can plan objectives be met in spite of the demands of unplanned activities?	___	___	___	___
14.	Is adequate time allocated in the plan for unplanned activities?	___	___	___	___
15.	Are present workplace health protection program activities directed more toward responding to regulatory issues than to the development and maintenance of a comprehensive program?	___	___	___	___
16.	Are adequate time and resources available to develop, implement, and maintain a comprehensive workplace health protection program?	___	___	___	___
17.	Is it possible to obtain the resources necessary to develop and implement those elements of a comprehensive workplace health protection program that are not a part of the present on-going program?	___	___	___	___

N/A - Not Applicable

WORKPLACE HEALTH PROTECTION PROGRAM ELEMENT CHECKLIST ON PLANNING/STEWARDSHIP

	Yes	No	Don't Know	N/A
18. Are additional resources provided, as needed, for addressing the workplace health issues associated with turnaround activities?	___	___	___	___
19. Is the future impact of regulations anticipated so that demands on resources will not result in planned activities being deferred or deleted from the plan?	___	___	___	___
20. Is there a procedure in place to communicate to management the potential impact of forthcoming regulations on the existing workplace health protection program?	___	___	___	___
21. Is there a procedure in place to periodically review workplace health protection program activities to ensure they are directed toward achieving the goals set by management?	___	___	___	___
22. Has management been provided (within the past three years) the results of a formal comprehensive review of the status of the workplace health protection program?	___	___	___	___

N/A - Not Applicable

WORKPLACE HEALTH PROTECTION PROGRAM ELEMENT
ON
RESOURCES

WORKPLACE HEALTH PROTECTION PROGRAM ELEMENT ON RESOURCES

OBJECTIVE

To provide information regarding resources that may be necessary for developing, implementing, and maintaining an effective workplace health protection program that will ensure a workplace without risks to health.

BACKGROUND

A variety of resources should be available to achieve the objective of safeguarding the health and well being of employees and the public that may be affected by the potential health hazards associated with operations. Personnel and equipment are needed to carry out the necessary activities to accomplish the overall goal of controlling workplace health hazards.

Regulatory requirements impact workplace health protection programs and require resources to carry out the activities necessary to achieve compliance. For example, regulations may incorporate requirements for hazard awareness training, monitoring, recordkeeping, protective equipment selection and fitting, and exposure control assessment. Thus, the impact of a regulation can result in the diversion of available resources from the goal of controlling all health hazards in the workplace to devoting essentially all resources to achieving compliance with a regulation which is concerned with only one substance.

The determination of benzene levels before entry into a confined space is an example of a regulatory requirement that has had an impact on the workplace health protection resources in the petroleum and petrochemical industries. It is due to the difficulty in measuring low benzene levels in the field before work can proceed (i.e., obtaining a permit). Direct reading methods are often not sufficiently sensitive for determining low benzene concentrations. It has been determined that an effective method to evaluate the benzene exposure hazard at low concentrations and identify appropriate protective equipment for personnel use is to obtain an air sample in a conditioned plastic bag for analysis on a gas chromatograph (GC) equipped with an appropriate detector. This approach results in significant expense and the need to train personnel in the specific GC unit use and data interpretation. Associated set-up, running, as well as maintenance and utility costs for this instrument can be high. Personnel need to obtain and analyze samples on request. Thus, they must be available on short notice to conduct such tests. The GC must be calibrated frequently, and the documentation of analytical results must be effective.

Non-achievement of goals can be a basis for justifying additional resources, particularly when it is due to "firefighting" and regulatory compliance activities. Responding to requests promptly can result in improved cooperation by personnel; enhancement of employee hazard awareness; better communications; and very likely, additional requests, which may well identify other concerns, with consequent improvement in workplace health protection. However, carrying out unplanned activities can affect the accomplishment of planned objectives.

The resource needs are ever changing. For example, if new toxicity information is developed on a material to which exposures may occur in a facility, and little or no monitoring data is available, a host of activities may be initiated which require that personnel interrupt their present work activities and address the new concern. This may involve conducting a literature search to identify an appropriate monitoring method, obtaining the instruments and sampling media to conduct a monitoring program, developing a sampling strategy, and carrying out the monitoring program.

A significant impact on resources may also result if there is a reduction of an existing exposure limit. A review of previous monitoring data may be carried out to determine where exposures have been detected and at what levels. While not always necessary, a review of the literature may be carried out to determine if the monitoring method used had been validated at the lower level of concern, and if not, whether there is an acceptable monitoring method available. As an example, if the proposed exposure limit for a substance is reduced to 25 percent of the present limit, it may require a monitoring method development effort to validate a sampling and analytical procedure that is effective at the lower concentration.

CONSIDERATIONS

The resources needed to develop, implement and maintain a workplace health protection program for a facility may include:

- Personnel
- Instrumentation
- Space
- Laboratory equipment
- Funding
- Plan
- Reference literature
- Miscellaneous items

The effectiveness of the program will depend upon obtaining the necessary resources, having the support of management to develop the program, a dedicated staff to implement and maintain it, and the cooperation of employees.

Personnel

The determination of personnel resource needs requires assessment by an industrial hygienist experienced in the facility's operations, recognition of the regulatory issues to be addressed, and an awareness of management's position on program needs.

Personnel resource needs will depend upon the policies of the organization; the exposure hazards encountered by employees; number of potentially exposed personnel; frequency and duration of their exposure to health hazards; training and monitoring requirements; the scope of the workplace health protection program elements to be implemented and maintained; regulatory issues impacting the program; and others. If there are a large number of acute and/or chronic toxicants to which employees may be exposed, more personnel could be needed to develop, implement, and maintain an effective comprehensive workplace health protection program. Where there are few exposure hazards associated with an operation, and there are infrequent contacts of short duration with the hazardous agents, then responsibility for the facility's workplace health protection program may be delegated to a technically trained individual who has access to a professional industrial hygienist for support and direction.

Those assigned workplace health protection program responsibility need to develop and maintain technical competence to effectively carry out their work. This is accomplished by attending technical courses/seminars/ conferences. The opportunity must be made available for personnel to do this.

Instrumentation

Instruments and accessories will be needed to evaluate employee exposure to chemical, biological and physical agents; determine the effectiveness of control measures; provide training to personnel; and for other purposes. The monitoring instruments must be properly maintained and calibrated, and appropriate records of this information retained.

The types and numbers of instruments necessary to carry out a monitoring program will depend on the hazards to be evaluated, size of the facility, number of potentially exposed personnel, instrument reliability, and other factors. These will have to be reviewed and a determination made regarding equipment needs. An industrial hygienist, knowledgeable in the facility's operations, could provide assistance in this regard.

Space

Office space will be necessary for personnel. Laboratory space should be provided with a bench and cabinets for the maintenance, calibration and storage of instruments. Filing space will be needed for monitoring records, survey results, instrument manuals, material safety data sheets/bulletins, the chemical inventory, radiation records, and copies of regulations. Book space will be needed for reference literature.

Laboratory Equipment

The laboratory equipment needs will depend upon the types of samples to be obtained. For example, a refrigerator may be needed if charcoal tube, charcoal badge, or other solid sorbent type samples are to be collected. A mechanically exhaust ventilated hood may be needed for the storage and mixing of some calibration and sorbent materials. An electrobalance would be appropriate if a significant number of aerosol samples are to be collected for the determination of the mass concentration of contaminants. In some facilities, the industrial hygiene function may have responsibility for evaluating bulk samples to determine the type and amount of asbestos in insulation and other materials. In addition, filter samples may be analyzed to determine the airborne asbestos fiber count. Microscopes are needed for these determinations.

Funding

The cost of the industrial hygiene activity will depend upon the number of personnel assigned to it; space required by the staff; scope of the program to be developed, implemented and maintained; the impact of regulations on the program; travel; analytical costs for samples collected; technical training of industrial hygiene function personnel; and other costs. A budget should be developed and submitted to management for approval as a part of the planning procedure.

Plan

An annual plan should be developed by the industrial hygiene group for review and approval by management. When approved, it becomes a resource of the industrial hygiene function. Accomplishments/shortcomings relative to planned activities should be reported to management periodically during the plan period.

Miscellaneous

A computing system, supported with appropriate software, should be available for maintaining monitoring records and data retrieval. Where appropriate, a vehicle (e.g., car, van or truck) should be available to the industrial hygiene staff for use in carrying out field activities.

Reference Literature

Technical literature should be available to the industrial hygiene function. A listing of basic texts, technical guides and journals that would be appropriate is shown in the Reference section.

REFERENCES

Textbooks

"Air Sampling Instruments for Evaluating Atmospheric Contaminants", Latest Edition (Cincinnati, OH: American Conference of Governmental Industrial Hygienists).

"Fundamentals of Industrial Hygiene", 3rd Edition (Chicago, IL: National Safety Council, 1988).

"Industrial Hygiene Aspects of Plant Operations - Process Flows", Vol. 1 (New York: Macmillan Publishing Co., 1982).

"Industrial Hygiene Aspects of Plant Operations - Unit Operations and Product Fabrication", Vol. 2 (New York: Macmillan Publishing Co., 1984).

"Industrial Hygiene Aspects of Plant Operations - Engineering Considerations in Equipment Selection, Layout and Building Design", Vol. 3 (New York: Macmillan Publishing Co., 1985).

"Industrial Noise and Hearing Conservation" (Chicago, IL: National Safety Council, 1975).

"In-Plant Practices for Job Related Health Hazards Control, Production Processes", Vol. 2 (New York: John Wiley & Sons, 1989).

"In-Plant Practices for Job Related Health Hazards Control, Engineering Aspects", Vol. 2 (New York: John Wiley & Sons, 1989).

"Industrial Ventilation - A Manual of Recommended Practice", Latest Edition (Lansing, MI: American Conference of Governmental Industrial Hygienists).

Miller, R. I., "The Industrial Hygiene Handbook for Safety Specialists" (Colombia, MD: Hanover Press, Inc., 1984).

"Occupational Diseases - A Guide to their Recognition", Revised Edition, DHEW (NIOSH) Publication No. 77-181 (Washington, D.C.: Superintendent of Documents, U.S. Government Printing Office, 1977).

"Patty's Industrial Hygiene and Toxicology - General Principles", Vol. 1, 3rd Edition (New York: John Wiley & Sons, 1978).

"Patty's Industrial Hygiene and Toxicology - Toxicology", Vol. 2A, 3rd Edition (New York: John Wiley & Sons, 1981).

"Patty's Industrial Hygiene and Toxicology - Toxicology", Vol. 2B, 3rd Edition (New York: John Wiley & Sons, 1981).

Patty's Industrial Hygiene and Toxicology - Toxicology", Vol. 2C, 3rd Edition (New York: John Wiley & Sons, 1982).

"Patty's Industrial Hygiene and Toxicology - Theory and Rationale of Industrial Hygiene Practice", Vol. 3A, 2nd Edition (New York: John Wiley & Sons, 1985).

Sax, N. I., "Dangerous Properties of Materials", Latest Edition (New York: Van Nostrand Reinhold Co.).

Shapiro, J., "Radiation Protection", 3rd Edition (Cambridge, MA: Harvard University Press, 1990).

"The Industrial Environment - Its Evaluation and Control", Stock No. 017-001-00396-4 (Washington, D.C.: Superintendent of Documents, U.S. Government Printing Office, 1973).

Technical Journals

American Industrial Hygiene Association Journal (Akron, OH: American Industrial Hygiene Association).

Applied Occupational & Environmental Hygiene, (Cincinnati, OH: Applied Industrial Hygiene, Inc., American Conference of Governmental Industrial Hygienists).

Safety & Health (Chicago: National Safety Council).

Technical Guides

American Industrial Hygiene Association Publications (Akron, OH: American Industrial Hygiene Association)

Biohazards Reference Manual, 1985.
Chemical Protective Clothing, Volumes 1 and 2, 1990.
Emergency Response Planning Guidelines (Sets 1, 2, 3, 4).

Engineering Field Reference Manual, 1984.
Ergonomic Guides
Fundamentals of Analytical Procedures in Industrial Hygiene, 1987.
Hygienic Guide Series
Noise and Hearing Conservation Manual, 4th Edition, 1986.
Quality Assurance Manual for Industrial Hygiene Chemistry, 1988.
Respiratory Protection: A Manual and Guideline, 1980.
Standards, Interpretation and Audit Criteria for Performance of Occupational Health Programs, 1976.

WORKPLACE HEALTH PROTECTION PROGRAM ELEMENT CHECKLIST ON RESOURCES

Personnel	Yes	No	Don't Know	N/A
1. Do personnel assigned workplace health protection program responsibility have available time to keep current on technical and regulatory issues?	___	___	___	___
2. Do personnel with workplace health protection program responsibility have access to external technical support?	___	___	___	___
3. Are external resources (corporate, consultant, etc.) utilized if needed?	___	___	___	___
4. Is the industrial hygiene function aware of all regulatory issues which impact workplace health protection issues?	___	___	___	___
5. Is the staff assigned workplace health protection program responsibility adequate to:				
• Support a comprehensive workplace health protection program?	___	___	___	___
• Support a program that is responsive to the approved plan and company policies?	___	___	___	___
• Respond in a timely manner to support a comprehensive program that meets company workplace health protection objectives and regulatory issues without disruption?	___	___	___	___
6. Are personnel assigned workplace health protection program activities provided adequate training to maintain skills and broaden the scope of their technical capabilities?	___	___	___	___
7. Are all regulatory and company required written workplace health protection program elements complete and up to date?	___	___	___	___

N/A - Not Applicable

WORKPLACE HEALTH PROTECTION PROGRAM ELEMENT CHECKLIST ON RESOURCES

	Yes	No	Don't Know	N/A
8. Are monitoring records available and complete?	___	___	___	___
9. Are MSDSs/MSDBs:				
. Available on all materials?	___	___	___	___
. The most recent version?	___	___	___	___
. Reviewed for completeness?	___	___	___	___
. Filed in an organized manner?	___	___	___	___
10. Is the follow-up procedure of previously made recommendations effective for ensuring problems are resolved?	___	___	___	___
11. Does the industrial hygiene staff:				
. Respond to medical department requests?	___	___	___	___
. Participate in investigations?	___	___	___	___
. Attend committee meetings (as appropriate)?	___	___	___	___
. Tour the facilities and get to know supervisors and other employees?	___	___	___	___
. Respond to employee requests?	___	___	___	___
. Respond to regulatory requirements?	___	___	___	___

Instrumentation

	Yes	No	Don't Know	N/A
12. Is available instrumentation:				
. Appropriate for assessing potential hazards associated with operations?	___	___	___	___
. Sufficient in number and type?	___	___	___	___
. Effective for the application?	___	___	___	___
. Adequately maintained?	___	___	___	___
. Periodically calibrated and results appropriately documented:				
- Personal pumps?	___	___	___	___
- Portable direct reading monitors for:				
. Radiation?	___	___	___	___
. Hydrocarbon vapor/gas?	___	___	___	___
. Hydrogen sulfide?	___	___	___	___
. Carbon monoxide?	___	___	___	___
. Ammonia?	___	___	___	___
. Chlorine?	___	___	___	___
. Sulfur dioxide?	___	___	___	___
. Other? (________________)	___	___	___	___

N/A - Not Applicable

WORKPLACE HEALTH PROTECTION PROGRAM ELEMENT CHECKLIST ON RESOURCES

	Yes	No	Don't Know	N/A
13. Are instrument operating manuals readily available?	___	___	___	___
14. Are detector tubes within the expiration period?	___	___	___	___
Space				
15. Is adequate space available for:				
• Personnel?	___	___	___	___
• Instrument storage?	___	___	___	___
• Maintaining and calibrating instruments?	___	___	___	___
• Reference literature?	___	___	___	___
• MSDSs/MSDBs?	___	___	___	___
• Survey reports?	___	___	___	___
• Monitoring results?	___	___	___	___
• Other? (____________________)	___	___	___	___
Equipment				
16. Is there a refrigerator available to store samples until shipped to the laboratory?	___	___	___	___
17. Is there a capability to ship refrigerated (cold) samples?	___	___	___	___
18. Is a gas chromatograph unit available to check samples for specific hydrocarbons (benzene or others) as part of the confined space entry, respirator, or other program?	___	___	___	___
19. Is a vehicle available to industrial hygiene function personnel for field work?	___	___	___	___
20. Is the electrobalance/balance effectively maintained and calibrated (i.e., yearly)?	___	___	___	___
21. Are appropriate microscopes and accessories available for each application (e.g., counting, sizing, identifying particles/fibers)?	___	___	___	___

N/A - Not Applicable

WORKPLACE HEALTH PROTECTION PROGRAM ELEMENT CHECKLIST ON RESOURCES

	Yes	No	Don't Know	N/A
22. Have personnel who use the microscope been trained to:				
. Identify fibers?	___	___	___	___
. Count fibers?	___	___	___	___
. Size particles?	___	___	___	___
Funding				
23. Is an annual budget developed for the industrial hygiene activity?	___	___	___	___
24. Is the funding provided for the industrial hygiene function (or others with program responsibility) adequate to carry out the planned program?	___	___	___	___
25. Is there a capability to obtain funding for activities that have not been budgeted for?	___	___	___	___
26. Are staff personnel supported in maintaining technical proficiency and/or certification?	___	___	___	___
Reference Literature				
27. Is there adequate technical literature available to those personnel with workplace health protection program responsibility?	___	___	___	___
28. Do personnel with workplace health protection responsibilities keep current on technical issues by reading technical journals?	___	___	___	___
Plan				
29. Is a workplace health protection plan developed annually?	___	___	___	___
30. Is the plan reviewed and approved by management?	___	___	___	___

N/A - Not Applicable

WORKPLACE HEALTH PROTECTION PROGRAM ELEMENT CHECKLIST ON RESOURCES

		Yes	No	Don't Know	N/A
31.	Is the plan stewarded to management?	___	___	___	___
32.	Do unplanned activities interfere with the accomplishment of planned activities?	___	___	___	___
33.	Is the planning procedure used to obtain additional resources (as needed)?	___	___	___	___
34.	Are workplace health protection program activities comprehensive and effective in addressing all potential hazards associated with operations?	___	___	___	___
35.	Are employee concerns related to workplace exposure hazards promptly addressed?	___	___	___	___
36.	Is the Medical Department satisfied that the support requested from personnel assigned workplace health protection program responsibility is prompt and thorough?	___	___	___	___

N/A - Not Applicable

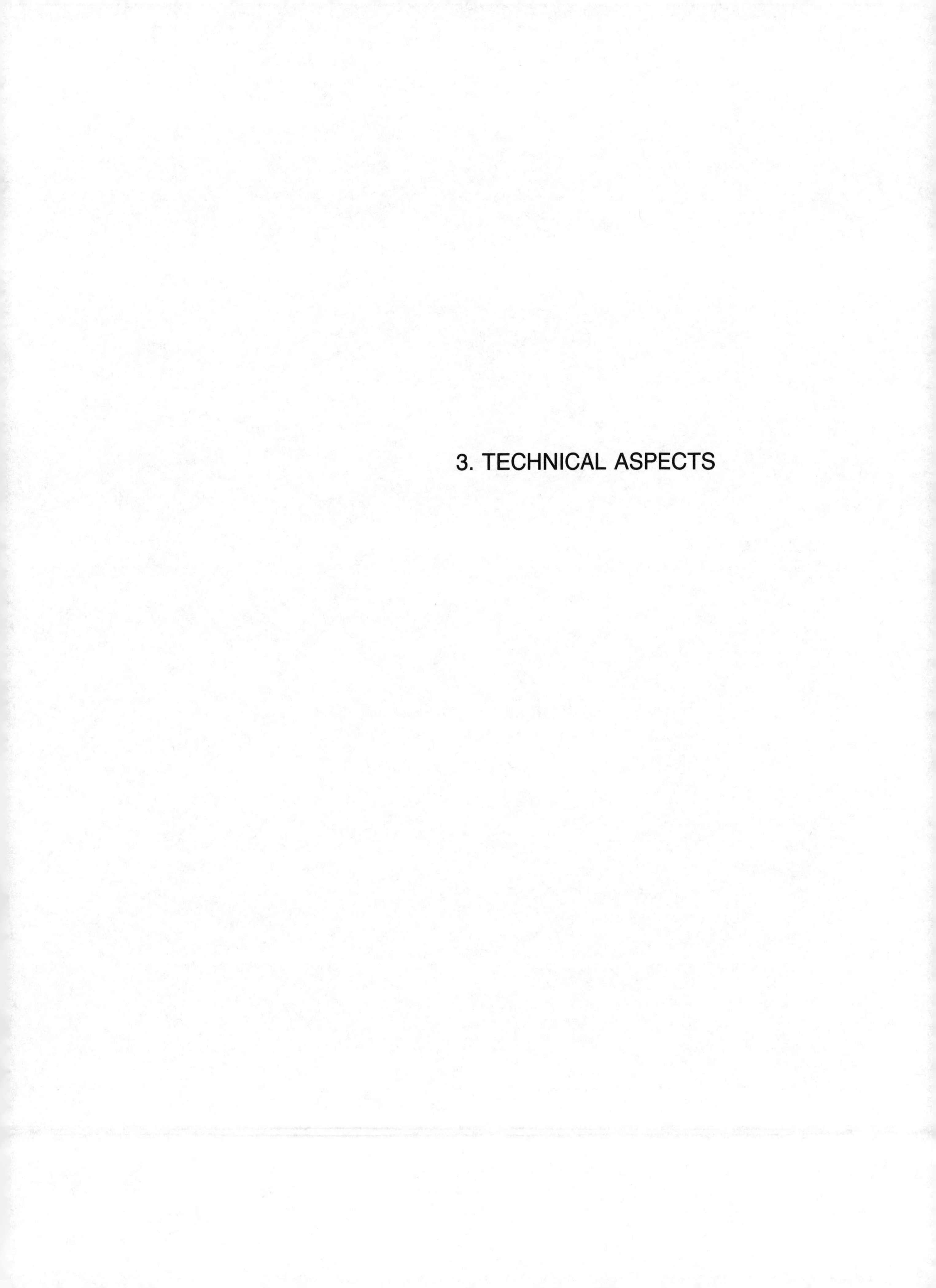

3. TECHNICAL ASPECTS

WORKPLACE HEALTH PROTECTION PROGRAM ELEMENT
ON
CONTINUOUS MONITORS

WORKPLACE HEALTH PROTECTION PROGRAM ELEMENT ON CONTINUOUS MONITORS

OBJECTIVE

To provide information for improving the reliability of continuous monitoring systems that are in operation for detecting contaminants of health concern and alerting personnel of potential health hazards.

BACKGROUND

Continuous air monitoring equipment is used to detect airborne contaminants in refineries, on offshore platforms, in petrochemical facilities, at petroleum marketing operations, gas plants, and other locations. These monitors are typically equipped with an audible and/or visual alarm for indicating when a pre-set concentration of the monitored substance is exceeded. They may be capable of shutting down a process, activating ventilation equipment, and alarming simultaneously at the detector location and in a control center. Some continuous monitors are installed for safety reasons, while others alert personnel of the existence of a potential health hazard.

Specific actions are triggered when an alarm is activated. This may include the requirement that personnel don respiratory protection, evacuate an area, shut down equipment or initiate other emergency procedures. The alarm should continue to be activated until manually reset.

The types of continuous monitors that may be in place for alerting personnel of a situation that could be hazardous to health include units for detecting hydrogen sulfide, sulfur dioxide, chlorine, carbon monoxide, oxygen deficiency, ionizing radiation, specific hydrocarbons (benzene, 1,3-butadiene, vinyl chloride, etc.), and others.

Units installed to detect flammable/combustible vapors and gases from a fire/explosion standpoint also provide information relative to health concern considerations when they alarm. For example: consider a continuous combustible gas/vapor monitor set to alarm at 25% of the lower explosive limit (LEL) of the substance of concern (e.g., acetone). If this monitor alarms in a potential acetone contaminated area, it would indicate that the airborne level of acetone vapor was at 25% of the lower explosive limit (LEL - 2.6%) for acetone. The part per million concentration of acetone at this level is 6,500 ppm (0.25 x 26,000 ppm = 6,500 ppm). This is well above the acceptable 8-hour exposure limit of 750 ppm for acetone. Even though the acetone level may be acceptable from a fire/explosion standpoint, it may be unacceptable from a health concern standpoint.

An unacceptable exposure situation may develop at a work location if a combustible gas/vapor type sensor is improperly located. They are typically positioned between the source(s) of release and the usual work location. In addition, continuous monitors are often positioned near points of contaminant release in order to provide a pre-warning long before the contaminant reaches a level of health concern at a work station.

It is a practice in industry to set the alarm level of continuous health monitors at, or slightly above, the acceptable 8-hour exposure limit for the substance being monitored. Thus, if the alarm is activated, an unacceptable exposure situation may develop. The alarm point decision must be based on the toxicity of the substance being monitored; sensor position; accuracy, precision, and reliability of the device; as well as the action that is triggered if/when an alarm is activated.

The positioning of the audible/visual alarm in the field is best if it is in the immediate vicinity of the sensor. This facilitates locating the source of contamination, as well as indicating to personnel where the release location is and the direction not to go if evacuating the area. Placing additional alarms (light/horn) at the entrance to a building, access way to a unit/area, and in a continuously occupied location (e.g., a control room) is also advisable.

CONSIDERATIONS

A detailed review of continuous monitor performance characteristics is essential before the purchase and installation of this type equipment. Cost should not be the prime determinant for deciding which unit to install. Important factors to consider in selection include toxicity (acute/chronic) of the substance; instrument reliability; sensor life; temperature effect; sensor drift; possible presence of interferences; ease of calibration; whether an active or passive system is needed; potential maintenance problems; availability of resources to service these devices; and others. Accuracy of response is not always the first concern in selection. Instrument reliability is often the most important consideration.

Personnel from various groups (industrial hygiene, safety, instrument, etc.) should be involved in the selection of continuous monitors, system design, location of sensors, alarm point setting, etc.

The monitors should alarm at the sensor location and in a central occupied location. Both visual and audible alarms should be provided. Dual level alarms are often desirable. In these, different actions are initiated based on the contaminant concentration which triggered the alarm.

Awareness of the presence of a continuous monitor in a work location is important for all personnel that enter the area. Employees and others working there should know what the alarm signals sound and look like, and be instructed in the procedures to follow if an alarm is activated. They should also know the proper evacuation route.

Continuous monitors need to be maintained and periodically calibrated (e.g., monthly or other documented frequency) to ensure continued effective operation. System response checks should be carried out more frequently (e.g., weekly or at least monthly). This involves a determination of whether each sensor in the system is operating properly and is normally carried out by a process operator or laboratory technician. Detailed calibrations are typically carried out by the instrument group.

Calibration gases must be on hand for the periodic response checks and the full calibration of continuous monitors. Calibrant gas concentration for periodic response checks should be about 20% above alarm settings. Calibrants for the full calibration should desirably be at 1/2, 1, and 2 times the low level alarm point. The higher concentration should be sufficient to trigger the high level alarm, if present.

Continuous monitors should be considered critical instruments. The frequency for periodic calibration and response checks should be specified in a written procedure. Records should be maintained of these checks. If there are carbon monoxide monitors on supplied air systems in the facility, they too should be considered critical instruments and be checked periodically.

Designated individuals should be given responsibility for coordinating the calibrations/response checks and ensuring that the groups assigned the task of carrying out these activities do it on schedule. Alarms should be activated periodically (with management concurrence) to ensure personnel respond to them properly.

REFERENCES

Smith, J. P. and S. A. Shulman, "An Evaluation of H_2S Continuous Monitors Using Metal Oxide Semiconductors", Applied Industrial Hygiene, July 1988, pp. 214-221.

American Petroleum Institute Recommended Practice No. 55 (API RP55), "Conducting Oil and Gas Production Operations Involving Hydrogen Sulfide" (Washington, D. C.: American Petroleum Institute, 1981).

WORKPLACE HEALTH PROTECTION PROGRAM ELEMENT CHECKLIST ON CONTINUOUS MONITORS

		Yes	No	Don't Know	N/A
1.	Are continuous air monitors that are in operation effective for alerting personnel of potentially hazardous situations?	___	___	___	___
2.	Are continuous air monitors considered critical instruments?	___	___	___	___
3.	Are there written procedures for carrying out maintenance, calibrations and response checks of continuous air monitors?	___	___	___	___
4.	Are there appropriate calibrants (gases/vapors) available for performing periodic response checks and calibrations of continuous air monitors?	___	___	___	___
5.	Has an individual been identified and given responsibility for assuring that the continuous air monitors are operating, in calibration, and being responded to as required?	___	___	___	___
6.	Are periodic (e.g., weekly/monthly) response checks carried out on all continuous air monitors in operation on the site?	___	___	___	___
7.	Are periodic (e.g., monthly or other documented frequency) calibrations performed on all continuous air monitors in operation on the site?	___	___	___	___
8.	Are written records maintained of the dates of response checks and calibrations, as well as of the results?	___	___	___	___
9.	Are records kept of maintenance problems that have occurred with each instrument in the system(s)?	___	___	___	___

N/A - Not Applicable

WORKPLACE HEALTH PROTECTION PROGRAM ELEMENT CHECKLIST ON CONTINUOUS MONITORS

		Yes	No	Don't Know	N/A
10.	Are personnel (employees, contractors) working in an area aware of:				
	. The presence of continuous air monitors?	___	___	___	___
	. The alarm signal system?	___	___	___	___
	. What to do if an alarm is activated?	___	___	___	___
11.	Do continuous air monitor alarms incorporate:				
	. A horn/siren or other audible signal?	___	___	___	___
	. A visual (light) signal?	___	___	___	___
	. Both audible and visual signals?	___	___	___	___
12.	Are continuous air monitor alarms activated:				
	. In the vicinity of the sensor?	___	___	___	___
	. In the immediate vicinity of the sensor?	___	___	___	___
	. Simultaneously in the unit/area and in an occupied location?	___	___	___	___
13.	Are the alarm signals visible and audible during normal operations, start-up, upsets, etc.?	___	___	___	___
14.	Have personnel been trained and made aware of how to respond to, evacuate from, and investigate a continuous monitor alarm situation?	___	___	___	___
15.	Is appropriate protective equipment and support available for responding to an alarm:				
	. Respiratory protection?	___	___	___	___
	. Portable monitors?	___	___	___	___
	. Protective clothing?	___	___	___	___
	. Back-up (standby) person?	___	___	___	___
16.	Are alarms operating properly:				
	. In the facility?	___	___	___	___
	. In an occupied location (e.g., control room)?	___	___	___	___
17.	Must alarms be reset manually?	___	___	___	___

N/A - Not Applicable

WORKPLACE HEALTH PROTECTION PROGRAM ELEMENT CHECKLIST ON CONTINUOUS MONITORS

	Yes	No	Don't Know	N/A
18. Is one individual/group given responsibility to assure the environment is acceptable for re-entry into the area after a continuous air monitor alarm has been activated?	___	___	___	___
19. Are personnel who may be required to respond to an alarm situation and to use respiratory protection:				
. Aware of the hazard?	___	___	___	___
. Included in a respirator medical surveillance program?	___	___	___	___
. Trained in the use of the protective equipment?	___	___	___	___
20. Have personnel who may use portable monitoring equipment in investigating an alarm situation been trained in its use and interpretation of the data obtained with it?	___	___	___	___
21. Are there procedures in place for alerting the surrounding community in the event of a spill/release that could impact it?	___	___	___	___
22. Is there communication with nearby facilities for alerting them of an emergency situation in your facility that could pose a health hazard to their personnel (and vice versa)?	___	___	___	___
23. Are the breathing air systems (cylinder filling, supplied air systems, etc.) equipped with a CO monitor?	___	___	___	___

N/A - Not Applicable

WORKPLACE HEALTH PROTECTION PROGRAM ELEMENT
ON
ERGONOMICS

WORKPLACE HEALTH PROTECTION PROGRAM ELEMENT ON ERGONOMICS

OBJECTIVE

To provide information for determining if cumulative trauma risk factors exist within a facility and whether the essential components of an ergonomics program should be in place to prevent cumulative trauma disorders.

BACKGROUND

Over the past several years there has been increased activity within industry to identify and modify workstations and operations that contribute to the development of cumulative trauma disorders (CTDs). CTDs are occupational illnesses that develop over time to affect the musculoskeletal and peripheral nervous systems. All joint structures are vulnerable to repeated trauma insults and each joint has its own family of cumulative trauma disorders. In general, the CTDs that occur are the result of inflamed tendons, tendon sheaths, or compressed nerves.

Guidelines have been developed for the prevention and control of the CTDs that have occurred in the meat packing industry. Many of the concepts developed in that program are adaptable for general industry use. These include:

- Management Commitment
- Employee Involvement
- Training and Education of Personnel
- Worksite Analysis to Identify Risk Factors
- Medical Management

Worksite analysis is a systematic and objective method for evaluating the ergonomic status of a facility based upon a review of each operation. The results of such analyses provide information for determining the need for a comprehensive ergonomics program or one which only addresses specific tasks.

CONSIDERATIONS

Worksite analyses should be carried out to identify operations at which CTD risk factors exist. It should be recognized that CTDs can affect any joint structure in the body but are most commonly associated with the hands, wrists, arms, elbows or shoulders. When a risk situation is observed, the body parts at risk should be identified and an assessment made of the likelihood for a cumulative trauma disorder to occur.

The incidence of CTDs have been correlated with combinations of the following risk factors:

- Forceful Exertions
- Frequent Exertions
- Extreme Joint Postures
- Localized Tissue Stress
- Vibration
- Low Temperatures

When combinations of these risk factors are observed among personnel performing a task, then a more detailed assessment or worksite analysis is warranted to determine the potential for the occurrence of a CTD. Work factors to be evaluated include whether:

- The operation requires a repeated pinch grip that has to be maintained for more than about 10 seconds
- The operation is performed several times per minute
- The task requires repeated forearm motion
- The individual performing the task is exposed to vibration
- The work is carried out in a cold environment
- The individual performing the work must keep the arms raised or winged out
- The individual performing the task works with the elbows extended behind the body.
- The individual must:
  - Kneel, squat, or stand on one leg
  - Remain in one position (sit or stand) continuously
  - Work with a part of the body pressed against an edge or surface
  - Use a part of the body as a striking tool
  - Bend the neck repeatedly or constantly
  - Bend the back more than 45 degrees
- Task involves performing a lift:
  - Of more than 20 pounds
  - Where there is poor footing
  - With a jerking motion
  - With a twisting motion
  - Where a good grip cannot be made
  - That also involves reaching for the object

If one or more of the above factors are observed, recommendations should be considered for modifying the task in order to minimize the potential effect of the cumulative trauma risk factors on the body part.

REFERENCES

"Work Practices Guide for Manual Lifting" [Cincinnati, OH: DHHS (NIOSH) Publication No. 81-122, National Institute for Occupational Safety and Health, 1981].

Keyserling, W. M., T. J. Armstrong and L. Punnett, "Ergonomic Job Analysis: A Structured Approach for Identifying Risk Factors Associated with Overexertion Injuries and Disorders", Applied Occupational and Environmental Hygiene, May 1991, pp. 353-363.

Tichauer, E. R., "Ergonomics", Patty's Industrial Hygiene and Toxicology, Vol. 1, 3rd Edition, Chapter 22 (New York: John Wiley & Sons, Inc., 1978), pp. 1059-1147.

Hertig, B. A., "Ergonomics", Fundamentals of Industrial Hygiene, 2nd Edition (Chicago, IL: National Safety Council, 1979), pp. 401-437.

"Ergonomic Guide Series" (Akron, OH: American Industrial Hygiene Association, Various dates of publication for specific guides in the series).

Hampel, G. A. and W. J. Henson, "Hand Vibration Isolation -A Study of Various Materials", Applied Occupational and Environmental Hygiene, December 1990, pp. 859-869.

Wasserman, D. E. "Reynaud's Phenomenon as it Relates to Hand-Tool Vibration in the Workplace", American Industrial Hygiene Association Journal, December 1985, pp. B 10-18.

Genaidy, A. M., et.al., "Improving Human Capabilities for Combined Manual Handling Tasks Through a Short and Intensive Physical Training Program", American Industrial Hygiene Association Journal, November 1990, pp. 610-614.

Keyserling, W. M., "Analysis of Manual Lifting Tasks: A Qualitative Alternative to the NIOSH Work Practices Guide", American Industrial Hygiene Association Journal, March 1989, pp. 165-173.

WORKPLACE HEALTH PROTECTION PROGRAM ELEMENT CHECKLIST ON ERGONOMICS

		Yes	No	Don't Know	N/A
1.	Has an effort been made to identify ergonomic problems/concerns in the facility?	___	___	___	___
2.	Is there a management commitment to the ergonomics program including:				
	. A written statement?	___	___	___	___
	. Written worksite analysis component?	___	___	___	___
	. Written worker training and education component?	___	___	___	___
	. A medical management component?	___	___	___	___
3.	Is there employee involvement in the ergonomics program, including an ergonomics committee with representative (facility-wide) constituency?	___	___	___	___
4.	Have injury/illness records been reviewed to determine CTD incidents?	___	___	___	___
5.	Is there an on-going effort to educate employees regarding ergonomic issues associated with tasks performed?	___	___	___	___
6.	Is a discussion on ergonomics and CTDs included in the facility's hazard communication program?	___	___	___	___
7.	Have the following been provided hazard awareness training on ergonomic issues and CTDs:				
	. Management?	___	___	___	___
	. Design engineers?	___	___	___	___
	. Health care providers?	___	___	___	___
	. Supervisors?	___	___	___	___
	. Other personnel? (____________)	___	___	___	___
8.	Have employees been surveyed regarding their opinion of CTD risk factors associated with tasks performed?	___	___	___	___

N/A - Not Applicable

WORKPLACE HEALTH PROTECTION PROGRAM ELEMENT CHECKLIST ON ERGONOMICS

		Yes	No	Don't Know	N/A
9.	Are CTDs that are observed/ diagnosed by medical brought to the attention of the industrial hygiene function for investigation?	___	___	___	___
10.	Are investigations (job analyses) carried out to determine the CTD cause and identify control methods?	___	___	___	___
11.	Does the job analysis team include:				
	. Employees?	___	___	___	___
	. Supervisor?	___	___	___	___
	. Health/safety professionals?	___	___	___	___
	. Physician?	___	___	___	___
	. Nurse?	___	___	___	___
	. Others?(________________)	___	___	___	___
12.	Have worksite analyses been carried out to determine if there are CTD risk factors associated with tasks being performed?	___	___	___	___
13.	Are cumulative trauma risk factors promptly assessed and corrected, where warranted, in order to minimize the potential for a disorder to occur?	___	___	___	___
14.	Do design engineers consider ergonomic issues in workplace design?	___	___	___	___
15.	Are work stations/areas designed to accommodate the person so that work can be done without stressing the body?	___	___	___	___
16.	Do supervisory personnel consider ergonomic issues when developing procedures/work practices?	___	___	___	___
17.	Are personnel trained to lift properly?	___	___	___	___

N/A - Not Applicable

WORKPLACE HEALTH PROTECTION PROGRAM ELEMENT CHECKLIST ON ERGONOMICS

		Yes	No	Don't Know	N/A
18.	Are materials stored so they are accessible without reaching, stretching, flexing the wrist or extending the arms over the head?	___	___	___	___
19.	Are tasks designed to be performed with minimum:				
	. Reaching?	___	___	___	___
	. Bending?	___	___	___	___
	. Twisting?	___	___	___	___
	. Lifting?	___	___	___	___
	. Pulling?	___	___	___	___
	. Pushing?	___	___	___	___
20.	Is there an effort to mechanize a repetitive task and eliminate the manual aspects of it?	___	___	___	___
21.	Have seats been provided to enable task performance from a seated rather than a standing position?	___	___	___	___
22.	Is work height acceptable (i.e., between mid-thigh and mid-chest)?	___	___	___	___
23.	Is the table, chair, and desk height acceptable?	___	___	___	___
24.	Are the tools that are in use ergonomically designed?	___	___	___	___
25.	Are ergonomic factors considered in hand tool selection?	___	___	___	___
26.	Have tool vibrations been reduced, where appropriate?	___	___	___	___
27.	Have efforts been made to eliminate frequent actuation of machines/tools using the:				
	. Finger?	___	___	___	___
	. Palm of the hand?	___	___	___	___
	. Foot?	___	___	___	___
	. Hip?	___	___	___	___

N/A - Not Applicable

WORKPLACE HEALTH PROTECTION PROGRAM ELEMENT CHECKLIST ON ERGONOMICS

		Yes	No	Don't Know	N/A
28.	Are tools properly maintained to accomplish a task with a minimum of effort?	___	___	___	___
29.	Is personal protective equipment available in different sizes/shapes to accommodate all personnel?	___	___	___	___
30.	Are breaks provided to personnel who are assigned to tasks requiring long periods of standing or sitting?	___	___	___	___
31.	Is lighting adequate so that shadows do not require personnel to stoop when performing a task?	___	___	___	___
32.	Has corrective action been considered for jobs/tasks with the following characteristics:				
	. Those requiring a pinch grip several times per minute and/or for longer than 10 seconds?	___	___	___	___
	. Those requiring repeated forearm motion?	___	___	___	___
	. Those which expose the individual to vibration?	___	___	___	___
	. Where there is exposure to a source of vibration in a cold environment?	___	___	___	___
	. Those requiring an individual to work with the arms raised, winged out, or with the elbows behind the body?	___	___	___	___
	. Lifting tasks where there is poor footing, a need to reach for an object, a twisting or jerking motion, or difficulty in achieving a good grip?	___	___	___	___
	. Those performed in a kneeling or squatting position?	___	___	___	___
	. Those which require bending the neck repeatedly or the back more than 45 degrees?	___	___	___	___

N/A - Not Applicable

WORKPLACE HEALTH PROTECTION PROGRAM ELEMENT CHECKLIST ON ERGONOMICS

		Yes	No	Don't Know	N/A
33.	Where appropriate, has there been an effort to reduce vehicular (trucks, industrial trucks, etc.) vibration for the operator?	___	___	___	___
34.	Are control panels properly designed to reduce visual stress?	___	___	___	___

N/A - Not Applicable

WORKPLACE HEALTH PROTECTION PROGRAM ELEMENT
ON
HEARING CONSERVATION

WORKPLACE HEALTH PROTECTION PROGRAM ELEMENT ON HEARING CONSERVATION

OBJECTIVE

To identify the potential for occupational hearing loss to occur among personnel working in petroleum and petrochemical operations and the means to reduce exposure risk through the implementation of an effective hearing conservation program.

BACKGROUND

Petroleum and petrochemical operations may generate sufficient noise such that personnel working in these facilities may experience hearing loss if adequate exposure controls are not implemented and hazard awareness training provided. Sources of noise include operating equipment (pumps, motors, compressors, etc.), flow control valves, movement of material through piping, furnace and boiler operations, steam leaks and others. The noise energy emitted by these sources can result in high noise levels (e.g., $\geq$ 85 dBA) in some areas of the facility. In order to reduce the potential for hearing loss to occur among employees, an effective hearing conservation program should be implemented.

CONSIDERATIONS

The purpose of a hearing conservation program is to prevent noise induced hearing loss among personnel exposed to noise in their work environment. A written program should be developed and implemented. It should include the measurement of area noise levels using appropriate equipment (e.g., a Type 1 or Type 2 sound level meter); identification of significant noise sources and a determination of the need for reduction of noise at the source; the assessment of employee exposure to noise; hazard awareness training; posting of high noise areas/sources; and providing personal hearing protection to noise exposed personnel, where appropriate.

Noise levels (e.g., dBA and dBC) should be determined throughout the facility with results presented on a plot plan grid of the area. Process units/areas with high noise levels (i.e., 85 dBA or more) should be identified, along with significant sources of noise. The measurements should be repeated periodically (every 2 - 3 years), as well as after new equipment is installed or significant process changes are made.

Units/areas with high noise levels should be posted with signs indicating the need for personnel to wear hearing protection. The hearing protection should be readily available to potentially exposed personnel and they should be trained in its use. It should be assured that the personal protective equipment provided is adequate to reduce noise exposure to an acceptable level.

Personal noise dosimetry studies should be carried out, if possible, to identify individuals, groups, job tasks, etc. that are excessively exposed to noise. A sufficient number of personnel should be monitored to provide an accurate profile of the exposure risk of each noise exposed group.

Personnel potentially exposed to noise should be provided hazard awareness training, as well as information on the effects of excessive exposure, areas where noise levels pose a hazard, and means available to them for reducing exposure risk (e.g., use of hearing protection).

Sources of noise that may be amenable to engineering control for reducing noise generation/emissions should be identified. Various groups (industrial hygiene, engineering, safety, environmental control, etc.) should work together to develop plans to control noise at the source, or through the implementation of other measures to reduce the noise levels to which employees are exposed.

An audiometric testing program should be established for the periodic evaluation of employee's hearing. Particular attention should be given to the test results of personnel who have been identified to be at high risk (from noise survey's and/or dosimeter studies). It should be assured that the audiometric test environment is acceptable (i.e., has a low background noise level), the audiometer is in calibration (i.e., meets applicable specifications), personnel to be tested have not been exposed to noise for about 14 hours prior to testing, and test results are reviewed promptly so that appropriate follow-up action can be taken. Use of personal hearing protection with proper noise attenuation is an acceptable means for controlling exposure to noise before an audiometric test is conducted. Where practical, personnel who conduct audiometric tests should be certified and provided periodic training.

Audiometric test results should be provided to the individuals tested, and the findings discussed relative to their use of exposure controls for preventing hearing loss. In addition, the audiometric test results should be reviewed periodically to determine the effectiveness of the hearing conservation program for preventing noise induced hearing loss.

REFERENCES

Michael, P. L., "Industrial Noise and Conservation of Hearing", Patty's Industrial Hygiene and Toxicology, Volume 1, 3rd Edition, General Principles (New York: John Wiley & Sons, 1978), pp. 275-357.

"Design Response of Weighting Networks for Acoustical Measurements", ANSI S1.42-1986 (New York: American National Standards Institute, 1986).

"Specification for Sound Level Meters", ANSI S1.4-1983 (New York: American National Standards Institute, 1983).

"Noise and Hearing Conservation", AIHA Monograph Series (Akron, OH: American Industrial Hygiene Association, 1981).

Olishifski, J. B., "Effective Hearing-Conservation Programs", Industrial Noise and Hearing Conservation (Chicago, IL: National Safety Council, 1975) pp. 705-795.

"Peterson, P. G. and E. E. Gross, Jr., "Handbook of Noise Measurement", 8th Edition (Concord, MA: GenRad, Inc., 1978).

Reynolds, J. L., J. H. Roysten and R. G. Pearson, "Hearing Conservation Programs (HCPs): The Effectiveness of One Company's HCP in a 12-Hr Work Shift Environment", American Industrial Hygiene Association Journal, August 1990, pp. 437-446.

Sataloff, R. T. and J. T. Sataloff, "Good Hearing - Conservation Program Detects Impairment and Saves Lives", Occupational Health and Safety, April 1991, pp. 46-50.

"Criteria For Permissible Ambient Noise During Audiometric Testing", ANSI S3.1-1977 (R1986) (New York: American National Standards Institute, 1986).

Cassli, J. G. and M. Y. Park, "Laboratory versus Field Attenuation" of Selected Hearing Protectors", Sound and Vibration, Oct. 1991, pp. 28-37.

"Specification for Personal Noise Dosimeter" ANSI S1.25-1978 (R1991) (New York: American National Standards Institute, 1991).

"Specifications for Audiometers", ANSI S3.6-1989 (New York: American National Standards Institute, 1989).

"Method for the Measurement of Real-Ear Attenuation of Hearing Protectors", ANSI S12.6-1984 (New York: American National Standards Institute, 1984).

**WORKPLACE HEALTH PROTECTION PROGRAM ELEMENT CHECKLIST
ON
HEARING CONSERVATION**

		Yes	No	Don't Know	N/A
1.	Is there a potential for employees to be exposed to an 8-hour time-weighted average noise level of 85 dBA or above?	___	___	___	___
2.	Is there a written hearing conservation program?	___	___	___	___
3.	Is the hearing conservation program in compliance with applicable laws and regulations?	___	___	___	___
4.	When appropriate, have exposure limits been established for work shifts different from that for an 8-hour workday?	___	___	___	___
5.	Are employees who have a potential time-weighted average exposure to noise of 85 dBA or greater included in the hearing conservation program?	___	___	___	___
6.	Is equipment available for carrying out:				
	• Area/source noise measurements? (Model/type of instrument(s)______)	___	___	___	___
	• Personal noise dosimetry studies? (Model/type of instrument(s)________)	___	___	___	___
7.	Does noise monitoring equipment meet or exceed applicable performance standards?	___	___	___	___
8.	Are means available to check the calibration of:				
	• Sound level meters?	___	___	___	___
	• Personal noise dosimeters?	___	___	___	___
9.	Are records of monitoring equipment calibrations effectively maintained?	___	___	___	___
10.	Do personnel use noise monitoring equipment properly (wind screen, proper incidence angle, avoid shielding, etc.)?	___	___	___	___

N/A - Not Applicable

WORKPLACE HEALTH PROTECTION PROGRAM ELEMENT CHECKLIST ON HEARING CONSERVATION

		Yes	No	Don't Know	N/A
11.	What data is obtained in evaluating noise exposure potential:				
	. dBA level?	___	___	___	___
	. dBC level?	___	___	___	___
	. Overall noise level?	___	___	___	___
	. Octave band analysis data?	___	___	___	___
12.	Are area noise levels determined throughout the facility every 2 to 3 years and when there is a significant change in equipment or the process?	___	___	___	___
13.	Are unit plot plans available showing noise levels on a grid pattern with noise level contours depicted?	___	___	___	___
14.	Are plot plans with noise level contours depicted on them available in control rooms?	___	___	___	___
15.	Have significant noise sources (e.g., $\geq$ 85 dBA) been identified in the facility?	___	___	___	___
16.	Is hearing protection:				
	. Readily available?	___	___	___	___
	. Of more than one type available?	___	___	___	___
	. Adequate, based on the noise exposure of the wearer?	___	___	___	___
17.	Are personnel provided training in the use of hearing protection?	___	___	___	___
18.	Is noise hazard awareness training provided to employees?	___	___	___	___
19.	Are all potential noise exposed personnel included in the noise hazard awareness training program?	___	___	___	___
20.	Is noise hazard awareness training of personnel repeated periodically?	___	___	___	___
21.	Are high noise areas posted to identify the exposure risk?	___	___	___	___

N/A - Not Applicable

WORKPLACE HEALTH PROTECTION PROGRAM ELEMENT CHECKLIST ON HEARING CONSERVATION

		Yes	No	Don't Know	N/A
22.	Are posted noise hazard warning signs effective (proper location, readily visible, message adequate, uniform, etc.)?	___	___	___	___
23.	Are noise monitoring results routinely transmitted to employees?	___	___	___	___
24.	Has a determination been made of the percentage of personnel who actually wear hearing protection in posted areas?	___	___	___	___
25.	Is audiometric testing of noise exposed personnel carried out periodically?	___	___	___	___
26.	Has the audiometric test environment been evaluated for background noise?	___	___	___	___
27.	Is the audiometric test environment acceptable?	___	___	___	___
28.	Is the audiometer checked biologically or acoustically before each day's use?	___	___	___	___
29.	Is an electronic ear available to check audiometer performance?	___	___	___	___
30.	Is the audiometer calibration checked acoustically on an annual basis?	___	___	___	___
31.	Was the audiometer within calibration when last tested?	___	___	___	___
32.	Are copies of audiometer calibration results available?	___	___	___	___
33.	Are personnel who conduct audiometric tests:				
	. Trained?	___	___	___	___
	. Certified, when appropriate?	___	___	___	___

N/A - Not Applicable

**WORKPLACE HEALTH PROTECTION PROGRAM ELEMENT CHECKLIST
ON
HEARING CONSERVATION**

		Yes	No	Don't Know	N/A
34.	Are test frequencies of 500, 1,000, 2,000, 3,000, 4,000, and 6,000 hertz included in the audiometer test procedure?	___	___	___	___
35.	Is there a practice in place to assure that personnel being given audiometric tests have been excluded from noise exposure for about 14 hours prior to their hearing test?	___	___	___	___
36.	Are audiometric test results reviewed promptly after testing to identify if a hearing shift has occurred?	___	___	___	___
37.	Is retesting done promptly if a hearing shift is noted in a periodic test?	___	___	___	___
38.	Is the industrial hygiene function and management kept informed of audiometric test results so that follow-up work (resurveys, evaluate hearing protection type and usage, etc.) can be performed, as necessary?	___	___	___	___
39.	Are follow-up recommendations/ requests made by the medical department carried out in a timely manner?	___	___	___	___
40.	Are periodic analyses carried out comparing audiometric test results with noise exposure findings?	___	___	___	___
41.	When feasible, have control measures (e.g., mufflers, sound absorbing walls, enclosures, vibration mounts, etc.) been implemented to reduce worker exposure to noise?	___	___	___	___

N/A - Not Applicable

WORKPLACE HEALTH PROTECTION PROGRAM ELEMENT CHECKLIST ON HEARING CONSERVATION

		Yes	No	Don't Know	N/A
42.	Has a plan been developed for installing engineering controls on significant noise sources?	___	___	___	___
43.	Are additions/deletions of personnel to/from the hearing conservation program promptly conveyed to the medical department?	___	___	___	___
44.	Are noise specifications incorporated into purchase orders for equipment which may be a significant source of noise?	___	___	___	___

N/A - Not Applicable

WORKPLACE HEALTH PROTECTION PROGRAM ELEMENT
ON
HEAT STRESS

WORKPLACE HEALTH PROTECTION PROGRAM ELEMENT ON HEAT STRESS

OBJECTIVE

To provide information for the assessment of heat stress and identify procedures and controls for the prevention of adverse health effects as a result of exposure to heat.

BACKGROUND

The combination of heat produced by the body and that of the environment in which a person works imposes a heat load on an individual. If the work rate is high, along with either the temperature or moisture content of the air, or both, a heat stress situation may arise. Excessive heat stress is more likely to develop among personnel when the ambient temperature and moisture content of the air are high. Generally, a dry bulb temperature above 90° F, a relative humidity of 80% or more with moderate to high dry bulb temperatures (e.g., greater than 85° F), or a radiant temperature (black globe) of 105° F or above may result in an excessive heat stress situation. The likelihood for excessive heat stress to occur increases if such conditions exist simultaneously with a moderate to heavy work rate (see Table 1).

Excessive heat stress situations may develop among personnel working in hot environments, such as inside or around furnaces, hot reactors, exchangers, etc., even though the ambient temperature and humidity may be acceptable. The likelihood for excessive heat stress to develop is increased when personnel have physical limitations or are not adequately acclimatized or adjusted to work in the hot environment. Thus, extra precautions and controls should be considered when unacclimatized or physically unfit personnel work in potentially stressful hot environments.

Excessive heat stress situations can result in heat stroke, heat exhaustion, cramps, or skin eruptions (prickly heat). Heat stroke involves an increase in body temperature with an absence of sweating. It is a life threatening situation requiring immediate cooling of the individual and prompt medical intervention. Heat exhaustion is a result of high volume blood flow to the skin and extremities with a consequent lack of blood supply to the internal organs. Periodic rest in a cool location can help prevent heat exhaustion from developing. Heat cramps involves muscle tightening or spasms in the extremities or the abdomen. Such cramps can be prevented by taking in an adequate amount of salt and water or an electrolyte type drink. Prickly heat is the result of the skin being wetted for extended periods, such as from sweating. This condition is more likely to develop where work is done in a hot humid environment. It can be prevented by the use of drying agents, putting on dry clothing, frequent showering, and/or keeping the more susceptible parts of the body dry.

There are several methods available to determine whether environmental conditions and the workload are likely to lead to an excessive heat stress situation among personnel. They are applicable to acclimatized workers. Included among these are:

Wet Bulb Globe Temperature Index (WBGT)

- Most widely used and accepted procedure and can be employed to assess exposures in the indoor and outdoor environment (see Table 2).

Heat Stress Index (HSI)

- Useful for identifying which factors to adjust/control for reducing exposure risk.

Effective Temperature Index (ET)

- Used for office type work situations. Typically used for assessing comfort issues rather than heat stress.

Black Globe Temperature (BGT)

- This method (e.g., use of the Botsball) is the simplest but the result is not equivalent to the WBGT value that would be obtained at the measurement location. Formulas have been developed to adjust the BGT reading to an equivalent WBGT value.

TABLE 1: WORKLOAD FOR VARIOUS TASKS

Nature of Work	Work Classification	Work Rate* kcal/hr	Work Rate* BTU/hr
Walking about with some lifting, pushing	Moderate	250	800
Sawing, operating impact wrench, building scaffold, etc.	Heavy	350	1200

* Approximate value

TABLE 2: FORMULAS FOR DETERMINING WET BULB GLOBE TEMPERATURE (WBGT)

For Indoor Application and Outdoors with No Solar Load:

WBGT = 0.7 x (Natural Wet Bulb Temperature)
+ 0.3 x (Black Globe Temperature)

For Outdoor Application with Solar Load:

WBGT = 0.7 x (Natural Wet Bulb Temperature)
+ 0.2 x (Black Globe Temperature)
+ 0.1 x (Dry Bulb Temperature)

CONSIDERATIONS

In order to prevent excessive heat stress among personnel, it is essential to recognize that heat stress is the result of a combination of several factors. Important among these are the work rate; time to complete a task; age, weight, and fitness of the worker; whether the individual is acclimatized; exposure controls in effect; availability of potable water or electrolyte fluid; work-rest ratio; and others. Controlling one or more of these can often reduce exposure risk to an acceptable level.

Many tasks that are carried out in petroleum and petrochemical operations require a moderate to heavy work rate. It is often not practical to reduce this factor. Thus, other means need to be considered for achieving heat stress exposure control. For example:

- Conducting medical evaluations of personnel to ensure their fitness to work in situations at which excessive heat stress exposure potential may be high.
- Establishing exposure guidelines for heat stressful situations.
- Providing training for personnel who may be involved in work situations at which heat stress may be excessive. Training should include information on heat stress symptoms and effects.
- Making provisions for personnel to become acclimatized before being assigned work where an excessive heat stress condition may develop.
- Assessing exposure risk of personnel doing tasks at which heat stress may be excessive.
- Adjusting the work-rest ratio, or having more than one individual be involved in performing a task.

- Identifying heat sources which, if controlled, could reduce exposure risk.

- Identifying those tasks for which heat stress exposure controls are necessary and implementing appropriate control measures.

- Modifying environmental conditions by reducing ambient temperature or water content of the air, increasing air velocity, or shielding sources of radiant heat.

- Ensuring that suitable rest areas are available to personnel who routinely work at tasks where the exposure potential for developing a heat disorder is significant.

- Referring individuals with heat stress symptoms to medical for evaluation.

- Providing appropriate personal protective equipment (aluminized suits, reflective face shields, vortex coolers, etc.) to personnel engaged in work where exposure to heat sources may result in an excessive heat stress situation.

REFERENCES

"Heating and Cooling for Man in Industry", 2nd Edition (Akron, OH: American Industrial Hygiene Association, 1975).

Mutchler, J. E., "Heat Stress: Its Effects, Measurement, and Control", Patty's Industrial Hygiene and Toxicology, 3rd Edition, Volume 1 (New York: John Wiley and Sons, Inc., 1977), pp. 927-992.

"Criteria for a Recommended Standard - Occupational Exposure to Hot Environment" (Cincinnati, OH: U.S. Department of Health, Education an Welfare, National Institute for Occupational Safety and Health, 1986).

Minard, D., "Physiology of Heat Stress", The Industrial Environment - Its Evaluation and Control (Cincinnati, OH: U.S. Department of Health, Education and Welfare, National Institute for Occupational Safety and Health, 1973).

Ramsey, J. D., "Practical Evaluation of Hot Working Areas", Professional Safety, February 1987, pp. 42-48.

Paull, J. M. and Rosenthal, F. S., "Heat Strain and Heat Stress for Workers Wearing Protective Suits at a Hazardous Waste Site", American Industrial Hygiene Association Journal, May 1987, pp. 458-463.

"An Improved Method for Monitoring Heat Stress in the Workplace", HEW Publication No. 75-1161 (Cincinnati, OH: U.S. Department of Health, Education and Welfare, National Institute for Occupational Safety and Health, 1975).

Beckett, W. S., et. al., "Heat Stress Associated With the Use of Vapor-Barrier Garments", Journal of Occupational Medicine, June 1986, pp. 411-414.

Kenney, W. L., et al, "Psychrometric Limits to Prolonged Work in Protective Clothing Ensembles", American Industrial Hygiene Association Journal, August 1988, pp. 390-395.

Gerson, V., "Heat Stress Strategies That Work", Safety and Health, May 1991, pp. 50-53.

Dernedde, E., "A Correlation of the Wet-Bulb Globe Temperature and Botsball Heat Stress Indexes for Industry", American Industrial Hygiene Association Journal, March 1992, pp. 169-174.

WORKPLACE HEALTH PROTECTION PROGRAM ELEMENT CHECKLIST ON HEAT STRESS

		Yes	No	Don't Know	N/A
1.	Have potential excessive heat stress exposures been identified in the work environment?	___	___	___	___
2.	Have potential excessive heat stress exposures been evaluated?	___	___	___	___
3.	Is there a means available for determining the magnitude of heat stress by the:				
	. Wet Bulb Globe Temperature (WBGT) Index method?	___	___	___	___
	. Heat Stress Index (HSI) method?	___	___	___	___
	. Black Globe Temperature (BGT) Index method?	___	___	___	___
	. Effective Temperature (ET) Index method?	___	___	___	___
4.	Have exposure guidelines been established for heat stressful situations?	___	___	___	___
5.	Are supervisory personnel aware of those tasks or work situations at which excessive heat stress may occur?	___	___	___	___
6.	Are acclimatization periods provided (e.g., after a sick leave, vacation, lay-off, etc.) for those workers involved in potential heat stress situations?	___	___	___	___
7.	Are personnel who may be exposed to heat stress situations:				
	. Identified?	___	___	___	___
	. Medically approved for the work?	___	___	___	___
	. Aware of the heat stress potential, as well as the symptoms of adverse effects and the measures to take to prevent such effects from developing?	___	___	___	___
	. Permitted to self-adjust their work rate?	___	___	___	___
	. Referred to medical for evaluation if heat stress symptoms develop?	___	___	___	___

N/A - Not Applicable

WORKPLACE HEALTH PROTECTION PROGRAM ELEMENT CHECKLIST ON HEAT STRESS

	Yes	No	Don't Know	N/A
8. Is cool water (or an electrolyte fluid) available to personnel within a reasonable distance from their work location?	___	___	___	___
9. Are employees encouraged to drink fluids during potential excessive heat stress periods?	___	___	___	___
10. Do personnel working at potentially excessive heat stress tasks have access to a rest area that is cooler than the work area?	___	___	___	___
11. When appropriate, are measures taken to reduce the potential for excessive heat stress to occur:				
. Provide for acclimatization of workers?	___	___	___	___
. Assure adequate cool water (or electrolyte drink) is available to workers?	___	___	___	___
. Implement engineering controls?	___	___	___	___
. Adjust work-rest ratio?	___	___	___	___
. Provide personal protective equipment?	___	___	___	___
. Limit frequency and/or duration of exposure?	___	___	___	___
. Allow for self-pacing of work by personnel performing task?	___	___	___	___
. Provide ventilation for personnel performing task?	___	___	___	___
. Schedule work during cooler part of the day?	___	___	___	___
12. Where appropriate, are the following available and used during situations when heat stress may be excessive:				
. Pedestal fan?	___	___	___	___
. General ventilation?	___	___	___	___
. Shielding?	___	___	___	___
. Insulation?	___	___	___	___
. Conditioned air?	___	___	___	___

N/A - Not Applicable

WORKPLACE HEALTH PROTECTION PROGRAM ELEMENT CHECKLIST ON HEAT STRESS

	Yes	No	Don't Know	N/A
. Protective clothing/equipment:				
- Gloves?	___	___	___	___
- Jacket?	___	___	___	___
- Suit?	___	___	___	___
- Eye protection?	___	___	___	___
- Face shield?	___	___	___	___
- Foot protection?	___	___	___	___
- Ice vest?	___	___	___	___
- Air cooling (vortex) system?	___	___	___	___
- Other? (______________)	___	___	___	___
13. Are steam or other hot lines in work areas effectively insulated?	___	___	___	___
14. Are steam leaks promptly controlled to reduce their heat contribution to the environment?	___	___	___	___
15. Are wooden sole shoes available for work where walking surfaces are hot (furnace firebox inspection/ maintenance, primary reformer top level, etc.)?	___	___	___	___
16. Is a maximum temperature established for entry into a hot environment (furnace firebox, etc.) for inspection, maintenance, etc.?	___	___	___	___

N/A - Not Applicable

WORKPLACE HEALTH PROTECTION PROGRAM ELEMENT
ON
IONIZING RADIATION

WORKPLACE HEALTH PROTECTION PROGRAM ELEMENT ON IONIZING RADIATION

OBJECTIVE

To identify procedures, work practices, and engineering controls that can be implemented to ensure personnel exposures to ionizing radiation are as low as reasonably achievable.

BACKGROUND

Sources of ionizing radiation are used in industrial operations for a variety of applications. These include metal inspection (identification of metal composition, radiography of weldments, etc.), as an analytical tool (X-ray diffraction unit, sulfur analyzer, etc.), in operations (thickness gauge, density gauge, level gauge, etc.), and for other applications. Use of radionuclides and radiation producing machines can result in personnel exposure to ionizing radiation that can have an adverse health effect if exposure is excessive. A generally accepted external radiation exposure limit for personnel who work with radiation sources (e.g., metal inspectors, radiographers, etc.) is 2 millirem (mrem) per hour (0.02 millisieverts per hour). The exposure limit for non-radiation workers is 10% of this value. Procedures and work practices should be established to limit exposures to no more than these levels and, where practical, to as low a level as is reasonably achievable.

Internal exposure limits are based on the specific isotope to which a person may be exposed, its solubility and target organ in the body. The potential for exposure to radiation sources that present an internal hazard is typically low in the petroleum and petrochemical industries. Where exposure potential is possible, appropriate controls must be implemented.

Fossil fuels, including crude oil and natural gas (and associated produced water), contain varying amounts of naturally occurring radioactive material (NORM). In their natural concentration in these fuels, the NORM typically present in fossil fuels (uranium, radium, and radon, along with its decay products) are not likely to cause a health concern. However, due to accumulations of residues around oil production or gas processing operations, NORM can become an environmental issue requiring clean-up if the concentration warrants. In the processing of crude oil and natural gas, NORM can gradually build up inside equipment, such as pumps, compressors, piping, manifolds, production tubing, separator drums, tanks, etc. Scale inhibitors can be used to reduce such deposition when this is a concern. Under typical operating conditions, excessive exposure to the NORM is unlikely to occur. However, when the equipment is opened, such as for maintenance, the accumulated scale can be released and result in an environmental concern, as well as a potential inhalation hazard if airborne particulates are released.

External radiation exposures can occur as a result of the build up of NORM scale containing radium. This can occur at production facilities where NORM contaminated water is produced, or where ethane and propane are separated from produced gases. If radon is present in natural gas it concentrates in the ethane and propane gas fractions and its decay products plate out on the inside of processing equipment, particularly pumps and compressors. Here they may, if sufficient amount accumulates, pose an external and internal exposure concern when the equipment is opened for maintenance/repair.

CONSIDERATIONS

To minimize exposure to ionizing radiation, the following procedures, work practices and control measures can be implemented.

General

- Develop a written radiation protection program.
- Establish the concept of limiting ionizing radiation exposures to as low as reasonably achievable.
- Provide training in basic radiological health for personnel potentially exposed to ionizing radiation.
- Give special consideration to pregnant women who may be exposed to ionizing radiation from the equipment/sources used in the facility. If at all possible, their exposure should be eliminated. As a minimum, their exposure should be monitored during the pregnancy period if there is a potential for any exposure to occur. It may be prudent to monitor all personnel who may routinely work with (or near) radiation sources.
- Identify a person with responsibility as the facility Radiation Safety Officer (RSO) and provide the training necessary to enable the individual to do an effective job.

- Identify a person(s) to function as the alternate RSO and provide appropriate training in radiation protection principles to enable this individual to do an effective job.

- Maintain and calibrate radiation monitoring instruments annually, or as required by regulation.
- Establish a procedure for the receipt of radioactive materials to ensure they have not been damaged in shipment such that they can present an exposure risk.

- Maintain an accurate and current inventory of radiation sources at the facility.

- Develop written procedures/plans for the safe use of ionizing radiation sources, as well as measures to take if an emergency arises that could affect the integrity of a source.

- Establish a recordkeeping system for sources of ionizing radiation on site. Included should be technical information on the device/source, purchasing and shipping records, registration/license, survey findings, wipe test results and other relevant data.

- Conduct a survey when a new, repaired, or relocated radiation source has been installed, and as required by license, where applicable.

- Ensure that appropriate hazard warning signs are posted.

- Establish a procedure to conduct periodic wipe tests of sealed sources of radioactive materials. A six month frequency is recommended.

- Assess exposure risk (via qualitative exposure assessments or work practice reviews) of personnel potentially exposed to ionizing radiation and provide personal radiation monitors (e.g., film badge or thermoluminescent dosimeter-TLD), when warranted, to determine individual exposures. Maintain records of results and review monitoring data periodically to ensure excessive exposures are not occurring. Investigate all occurrences of excessive exposure.

- A remote, unoccupied area should be established for the storage of radioactive materials that are to be installed, disposed, returned to the vendor, etc. Appropriate signs and access restrictions should be provided at this location.

Industrial Radiography

- Establish a procedure to ensure the RSO is informed when contract radiographic work is to be done and when the radiographer comes into the facility to perform the work.

- Ensure that the radiographer is licensed; knows the facility's emergency signals and what to do if an emergency arises; has an operable and recently calibrated radiation meter for monitoring the work area; wears a radiation exposure monitor; has the necessary rope and signs to establish a restricted access area around the work location; the radioactive source has been wipe tested (to assure no leakage) within the past 6 months; source labeling, records, and storage facility are acceptable; and has an emergency procedure to follow in the event of source loss, malfunction of equipment, etc.

- Ensure the radiographer's source holder is properly tagged or labeled with a radiation symbol, identification of the source material, its strength, and measurement date. In addition, records should be available on instrument calibration, leak tests, source use log, and personal exposure monitoring results. A written emergency procedure should be available.

- Ensure that the radiographer obtains the required permission for the work; clears personnel from the area; establishes a 2 mrem/hr (20 microsieverts) restricted area; posts the appropriate signs; monitors the area while the work is being performed; and assures that the exposure limit is not exceeded in potentially occupied areas.

- Ensure the radiographer uses a meter to indicate that the source has been effectively retracted into the camera, and the source holder is promptly (on completion of work) returned to a properly identified storage area (truck/room/building) that can be locked.

X-Ray Producing Equipment

- Ensure there is a current registration/license/permit for use of the equipment (as required by regulation); that each device has the appropriate caution radiation sign, symbol and label; and has effective interlocks which prevent access to internals while energized.

- Ensure personnel wear a radiation exposure monitor when operating the device (if warranted), and exposure records are properly maintained.

- Conduct a radiation survey of the radiation producing device and the surrounding area periodically (e.g., annually) to ensure stray radiation levels are acceptable.

Lab Analyzers (with Radioactive Source)

- Arrange for the source to be registered/licensed (as required to comply with regulations) and establish a wipe test procedure to check for source leakage.

- Ensure that a radiation symbol and appropriate label are on each instrument. The label should indicate the radioactive material, its strength, and measurement date.

- Ensure interlocks work and a procedure exists, when appropriate, to dispose of the device/material.

Process Analyzers (Including Gauges)

- Arrange for each device to be registered/licensed (as necessary to comply with regulations) and establish a wipe test procedure.

- Ensure a radiation symbol and caution statement are on on each device and the source is properly installed and labeled (indicating material, strength, and measurement date).

- Require that a survey be conducted immediately after a process analyzer has been installed to determine the need for caution signs, access restrictions, etc.

- Develop, with process personnel, written plans to be followed if an emergency arises in the unit/area where a source is in use. A key to lock the source shutter closed should be available to the RSO and the operator familiar with the emergency procedure.

- Close the source shutter while wipe testing or doing maintenance on the device and, if practical, when work is being performed near the source which could result in excessive radiation exposure.

- Require that a radiation survey be conducted if anyone is to enter a vessel on which a source is located. Before entry is permitted the source holder should be locked in the "shutter closed" position.

- Develop a procedure to dispose of radiation sources when no longer useful. This generally involves sending the instrument back to the supplier or having the supplier change out the source and dispose of the depleted one.

Medical X-Ray Machine

- Ensure medical X-ray machines are registered and licensed with the appropriate authority (as required), are properly labeled, and located such that the potential for exposure to stray radiation associated with their use is as low as reasonably achievable.

- Ensure appropriate beam collimators and filters are used in the unit during radiographic procedures and that a lead apron and gloves are available as needed (i.e., for patient gonadal shielding, fluoroscope applications, etc).

- Ensure that the operator is in a low exposure location during the medical radiography procedure.

- Provide, as appropriate, lead shielding for the X-ray room (walls, ceiling, floor, and operator's position) and a leaded glass viewing port at the operator's position.

- Evaluate the radiation exposure hazard potential associated with the use of the medical X-ray machine. This should be done periodically by the RSO.

- Provide the medical X-ray machine operator radiation protection training, a film badge or TLD, and maintain records of exposure results.

NATURAL OCCURRING RADIOACTIVE MATERIAL

Surveys should be carried out in facilities where there is a potential for exposure to Naturally Occurring Radioactive Material (NORM). These include at oil production operations; produced water handling facilities; gas separation facilities; LPG recovery, storage, and pumping facilities; and others. Measurements should be made to identify equipment that may pose an external and/or internal radiation exposure hazard to personnel or adversely affect the environment. This can be accomplished using a sensitive gamma radiation detector or sensitive dose-rate meter to make readings on the surface of suspect equipment/piping. Gamma radiation levels 2-3 times natural background [average about 0.01 milliroentgens per hour (0.1 microsievert per hour)] may indicate the presence of NORM at levels that could present an environmental or personnel exposure concern if the NORM is released into the environment or inhaled/ingested by personnel.

When equipment which exhibits internal contamination is to be taken out of service for maintenance or repair, precautions should be taken to reduce the potential radiation exposure hazard or contamination of the environment. Depending upon the nature of the concern (environmental, internal/external radiation exposure), this may include: recovery and appropriate disposal of scale/residue at production/processing sites, where warranted; draining and flushing of potentially contaminated equipment after allowing it to remain idle (not less than 4 hours) to reduce the external and internal radiation exposure hazard; using appropriate personal protective equipment (disposable gloves, respirator, etc.) and ventilation to reduce the inhalation/ingestion hazard (e.g., during brushing, grinding, or welding of contaminated surfaces); and collection of scale/residue/washings, when necessary, for appropriate disposal.

REFERENCES

Berlin, R. E. and C. C. Stanton, "Radioactive Waste Management" (New York: John Wiley & Sons, Inc., 1989).

Shapiro, J., "Radiation Protection", 3rd Ed. (Cambridge, MA: Harvard University Press, 1990).

"Basic Radiation Protection Criteria", NCRP Report No. 39, (Bethesda, MD: National Council on Radiation Protection and Measurements, 1971).

"Health Effects of Exposure to Low Levels of Ionizing Radiation: BIER V" (Washington, D. C.: National Academy Press, 1990).

"Radiological Health Handbook" (Rockville, MD: U.S. Department of Health, Education and Welfare, 1970).

"Health Physics and Radiological Health Handbook" (Olney, MD: Nuclear Lectern Associates, Inc. 1984).

Gesell, T. F., "Occupational Radiation Exposure Due to Rn-222 in Natural Gas and Natural Products", Health Physics Journal, November, 1975, pp. 681-687.

Morgan, K. Z., and J. E. Turner, "Principles of Radiation" (New York: John Wiley & Sons, Inc., 1967).

"Recommendations of the International Commission on Radiological Protection", ICRP Publication 26 (Oxford, England: Pergamon Press, 1977).

"Radiation Protection Instrumentation Test and Calibration", ANSI N323-1978 (New York: American National Standards Institute, 1978).

Summerlin, J., Jr. and H. M. Prichard, "Radiological Health Implications of Lead-210 and Polonium-210 Accumulations in LPG Refineries", American Industrial Hygiene Association Journal, April 1953, pp. 202-205.

Wilkening, G. M., "Ionizing Radiation", Patty's Industrial Hygiene and Toxicology, Volume 2, 3rd Ed. (New York: John Wiley & Sons, Inc., 1978), pp. 441-512.
"SI Units in Radiation Protection and Measurements", NCRP Report No. 82, (Bethesda, MD: National Council on Radiation Protection and Measurements, 1985).

"Bulletin on Management of Naturally Occurring Radioactive Matrials (NORM) in Oil & Gas Production.", API Bulletin E2, American Petroleum Institute, Washington, D.C., April 1992.

WORKPLACE HEALTH PROTECTION PROGRAM ELEMENT CHECKLIST ON IONIZING RADIATION

		Yes	No	Don't Know	N/A
1.	Is there a regulation that must be complied with relative to the possession or use of the radiation sources that are in the facility?	___	___	___	___
2.	Are sources of ionizing radiation registered/licensed as required?	___	___	___	___
3.	Has an individual been identified as the site Radiation Safety Officer (RSO)?	___	___	___	___
4.	Has the RSO and an alternate been provided training appropriate to their responsibilities?	___	___	___	___
5.	Has a written radiation procedure/ program been developed for the site?	___	___	___	___
6.	Have procedures been developed for the safe operation of each radiation source and each radiation producing device?	___	___	___	___
7.	Is a radiation survey conducted promptly after a radiation source is put into operation or is relocated?	___	___	___	___
8.	Have written plans been developed regarding steps to take if an emergency arises that could affect the integrity of a radiation source?	___	___	___	___
9.	Are copies of applicable radiation regulations available in the facility?	___	___	___	___
10.	Is there a written procedure for receipt of sources of ionizing radiation?	___	___	___	___
11.	Is there a calibrated radiation survey meter available for use in the facility?	___	___	___	___

N/A - Not Applicable

WORKPLACE HEALTH PROTECTION PROGRAM ELEMENT CHECKLIST ON IONIZING RADIATION

		Yes	No	Don't Know	N/A
12.	Is the radiation meter periodically calibrated (annually is recommended)?	___	___	___	___
13.	Is a check source available to ensure the meter is working properly?	___	___	___	___
14.	Are radiation symbols and caution signs present on:				
	. X-ray units?	___	___	___	___
	. Radioactive sources?	___	___	___	___
15.	Is a warning light energized to indicate when an X-ray unit is on?	___	___	___	___
16.	Are interlocks effective on X-ray units and other equipment where installed?	___	___	___	___
17.	Are radiation sources properly identified:				
	. Appropriate symbol?	___	___	___	___
	. Material identified?	___	___	___	___
	. Activity indicated?	___	___	___	___
	. Measurement date?	___	___	___	___
	. Signs in area?	___	___	___	___
18.	Are fire and emergency response groups aware of the location of each radiation source in the facility?	___	___	___	___
19.	Have personnel who are potentially exposed to ionizing radiation been informed of the hazard and health effects of excessive exposure?	___	___	___	___
20.	Are sealed radiation sources leak tested periodically (e.g., every 6 months)?	___	___	___	___
21.	Are records of radiation source registrations, licenses, inventory, wipe tests, calibration of equipment, etc., up to date?	___	___	___	___

N/A - Not Applicable

WORKPLACE HEALTH PROTECTION PROGRAM ELEMENT CHECKLIST ON IONIZING RADIATION

		Yes	No	Don't Know	N/A
22.	Do process operators know the location of radiation sources that are present in their unit, and where the key is to close (lock) the instrument shutter when appropriate to do so?	___	___	___	___
23.	Are personnel who work with radiation sources provided with:				
	. Pocket dosimeter?	___	___	___	___
	. Film badge?	___	___	___	___
	. Thermoluminescent dosimeter (TLD)?	___	___	___	___
24.	If film badges or TLDs are provided to facility personnel, is there a procedure to check vendor (processor) performance (e.g., use of control badges, replicates, etc.)?	___	___	___	___
25.	Are radiation exposure results and survey findings effectively maintained?	___	___	___	___
26.	Are personal exposure results reviewed periodically?	___	___	___	___
27.	Is there a written procedure in place to investigate high radiation exposures (e.g., greater than the applicable permissible exposure limit)?	___	___	___	___
28.	Have significant exposures to ionizing radiation been prevented (e.g., >25% of the permissible limit)?	___	___	___	___
29.	Are special precautions taken to limit exposure of pregnant women to radiation at levels as low as reasonably achievable?	___	___	___	___
30.	Are there regulations limiting exposure of pregnant women and the fetus to ionizing radiation?	___	___	___	___

N/A - Not Applicable

WORKPLACE HEALTH PROTECTION PROGRAM ELEMENT CHECKLIST ON IONIZING RADIATION

	Yes	No	Don't Know	N/A
31. Is there a written procedure for the disposal of old (depleted) radiation sources?	___	___	___	___
32. Are discussions held with the contract radiographers (to ensure they have equipment, license, signs, ropes, etc.) before they are permitted to carry out work on site?	___	___	___	___
33. Is the RSO, or alternate RSO, informed when a contract radiographer comes into the facility to perform work?	___	___	___	___
34. Does the contract radiographer:				
. Know the emergency procedures (contractor's and site's)?	___	___	___	___
. Have an operable, calibrated radiation meter?	___	___	___	___
. Have a sealed radiation source that has been checked for leakage within the past six months?	___	___	___	___
. Wear a film badge or other personal radiation monitor?	___	___	___	___
. Have a proper work permit?	___	___	___	___
. Properly position the rope barrier and post appropriate signs?	___	___	___	___
. Ensure all personnel are out of the restricted access area?	___	___	___	___
. Monitor the rope barrier to ensure the radiation level is not greater than 2 mrem per hour?	___	___	___	___
. Monitor the work area during the procedure?	___	___	___	___
35. Does the radiographer:				
. Use a radiation meter to determine that the source has been retracted into the holder?	___	___	___	___
. Remove the source from the area promptly after work is completed?	___	___	___	___

N/A - Not Applicable

WORKPLACE HEALTH PROTECTION PROGRAM ELEMENT CHECKLIST ON IONIZING RADIATION

	Yes	No	Don't Know	N/A
. Place the source in storage and lock the containment (truck/ storage shed/etc.)? . Inform the area "owner" when work is completed?	___	___	___	___
36. Does the radiographer's vehicle or site storage area have the proper hazard warnings (radiation symbol and caution sign)?	___	___	___	___
37. Has the radiographer's performance been checked in the past six months?	___	___	___	___
38. Are records maintained of the disposal of the facility's radiation sources (radionuclides, X-ray units)?	___	___	___	___
39. Is the ionizing radiation program in compliance with all aspects of applicable regulations?	___	___	___	___
40. Are all requirements of the facility's radioactive material license being met?	___	___	___	___
41. Have the facility emergency response groups been informed of the location of each radioactive source?	___	___	___	___
42. Have determinations been made for identifying equipment which may contain naturally occurring radioactive material (NORM)?	___	___	___	___
43. Are scale inhibitors used to reduce NORM deposition in equipment?	___	___	___	___

N/A - Not Applicable

WORKPLACE HEALTH PROTECTION PROGRAM ELEMENT CHECKLIST ON IONIZING RADIATION

		Yes	No	Don't Know	N/A
44.	If warranted, has a written procedure been developed for taking NORM contaminated equipment out of service and conducting maintenance or repairs?	___	___	___	___
45.	Where warranted, has a written procedure been developed for work which could result in NORM contamination of the environment?	___	___	___	___
46.	Are exposure controls adequate for minimizing employee exposure to NORM during:				
	. Routine operations?	___	___	___	___
	. Opening equipment for maintenance/repair?	___	___	___	___
	. Maintenance/repair of potentially contaminated equipment?	___	___	___	___
	. Disposal of NORM containing scale/residue?	___	___	___	___
47.	Is NORM contaminated equipment/land effectively decontaminated before release for unrestricted use?	___	___	___	___
48.	Are wet methods used to clean NORM contaminated equipment (as opposed to dry methods)?	___	___	___	___
49.	Are measurements made to determine the radiation level in potential NORM contaminated vessels before entry is permitted?	___	___	___	___
50.	Is water that is used to clean NORM contaminated equipment properly handled and filtered to remove contaminants, when appropriate?	___	___	___	___
51.	Is NORM containing sludge/scale stored in properly identified sealed containers?	___	___	___	___

N/A - Not Applicable

WORKPLACE HEALTH PROTECTION PROGRAM ELEMENT
ON
LIGHTING

WORKPLACE HEALTH PROTECTION PROGRAM ELEMENT ON LIGHTING

OBJECTIVE

To provide general information for developing, implementing and maintaining an acceptable lighting program.

BACKGROUND

The purpose for providing proper illumination is to enable personnel to work effectively without risk of injury due to inadequate light, glare, or other factors related to lighting. The primary physiologic benefit for establishing and maintaining proper lighting is that it reduces the effort to see. This results in reduced eye stress, visual fatigue, and headaches.

The feeling of visual comfort is an appropriate criterion on which to establish acceptable levels of lighting. The Illuminating Engineering Society has published tables which specify lighting levels for seeing in the horizontal work plane for various work tasks and operations. These are applicable for young adults with normal vision. Older individuals, and those with some loss of vision, will need higher levels of illumination to work effectively without risk of injury.

The visibility required to perform a task or observe an object is determined by its size, time for observation, detail to be observed, color, contrast, and brightness. Each factor is dependent upon the others, such that a deficiency in one may be compensated by increasing one or more of the others to achieve the same visual effect.

The quantity of illumination is measured in lumens per square foot (footcandles) or as lux (lumens per square meter). One footcandle is equivalent to approximately 10 lux. Recommended illumination levels for various tasks that have been established by the Illuminating Engineering Society are the minimums to be provided. Higher initial lighting levels must be provided to establish and maintain these levels of illumination since there will be a loss in lamp output and light transmission/reflectance with time. Where necessary, supplementary or direct lighting may be used in combination with general lighting to achieve recommended levels. The illumination levels listed in the tables of the referenced document are those which should be specified or recommended for the tasks/operations being carried out.

The quality of lighting pertains to the distribution of illumination in the visual environment. It is related to the amount of illumination on the task, glare (direct and reflected), distribution of light, as well as the color and brightness ratios of the surroundings. Lighting of very poor quality is often easily recognizable.

CONSIDERATIONS

To develop, implement and maintain an acceptable lighting program, there must be recognition and acceptance of the need to accomplish this. Many believe that if some lighting is provided in an area or at an operation, it will be adequate. There is often little effort to achieve and maintain the recommended level of illumination on a task or to consider quality factors such as glare, color of lighting, etc. Thus, it is necessary to "sell" the need for a lighting program.

After lighting fixtures have been installed there is often little effort to maintain recommended lighting levels. Illumination continues to decrease at work positions until a general relamping effort is necessary. In many situations lamps are replaced but fixtures are not cleaned.

The basic elements for achieving a good lighting program are to design the system properly; maintain it; periodically, make measurements to determine the adequacy of illumination levels and the existence of problems related to the quality of lighting (e.g., existence of glare sources, excessive brightness ratios, etc.); and identify deficiencies so that corrective action can be taken. Following is a list of factors to consider for developing and maintaining a good lighting program:

- Facility lighting should be designed to provide established in-service illumination levels at the work locations under consideration. The in-service level is considerably below the design level. This is due to lamp aging; soiling of fixtures, walls and ceilings; environmental factors; etc.

- Consideration should be given to the potential problems associated with relamping. These should be evaluated at the design stage and fixtures located so that cleaning and lamp replacement will not be difficult.

- Uniform distribution of illumination should be provided in continuously occupied areas. Although this is not essential for non-continuously occupied areas, the variation in lighting levels should be limited so that it is not objectionable.

- A periodic lighting maintenance program should be established to ensure lighting levels continue to be adequate. It should include lamp replacement and fixture cleaning. Maintenance frequency will be determined through experience (life of lamps, soiling, etc.) and the results of periodic lighting surveys in which lighting measurements are carried out, surface reflectance evaluated, and the quality of the illumination assessed relative to the task being carried out. In areas where oil mists, solids or corrosive gases are released into the air, maintenance needs will be more frequent. This will likely be true also for those areas/operations where structural vibrations exist. Special fixtures/equipment should be selected for such locations.

- Emergency lighting should be provided in selected areas to enable continued work, where necessary; the orderly egress of personnel in the event of an evacuation; rescue of individuals; or the shutdown of an operation. The emergency lighting level and location of fixtures should be evaluated to assure they provide the necessary illumination for the purpose intended. Records of such tests/inspections should be maintained.

REFERENCES

"IES Lighting Handbook", Fifth Edition (New York: Illuminating Engineering Society, 1972).

"American National Standard Practice for Industrial Lighting", ANSI/IES RP7-1983 (New York: American National Standards Institute, 1983).

"American National Standard for Office Lighting", ANSI/IES RPI-1982 (New York: American National Standards Institute, 1982).

Crouch, C. L., "Lighting for Seeing", Patty's Industrial Hygiene and Toxicology, Vol. 1, 3rd Edition (New York: John Wiley & Sons, Inc., 1978), pp. 513-594.

Kaufman, J. E., "Illumination", The Industrial Environment - Its Evaluation and Control (Cincinnati, OH: American Conference of Governmental Industrial Hygienists, 1973), pp. 349-356.

Largent, E. J. and J. B. Olishifski, "Fundamentals of Industrial Hygiene", 2nd Edition (Chicago, IL: National Safety Council, 1979), pp. 311-367.

WORKPLACE HEALTH PROTECTION PROGRAM ELEMENT CHECKLIST ON LIGHTING

		Yes	No	Don't Know	N/A
1.	Are applicable lighting specifications adhered to in the design and evaluation of lighting?	___	___	___	___
2.	Is there a facility-wide lighting maintenance program?	___	___	___	___
3.	Are periodic lighting surveys carried out in:				
	. Control rooms?	___	___	___	___
	. Laboratory?	___	___	___	___
	. Maintenance shops?	___	___	___	___
	. Manufacturing areas?	___	___	___	___
	. Offices?	___	___	___	___
	. Off-site locations?	___	___	___	___
	. Powerhouse?	___	___	___	___
	. Process areas?	___	___	___	___
	. Warehouse?	___	___	___	___
	. Other locations (________________)?	___	___	___	___
4.	Are the results of lighting surveys retained for future reference?	___	___	___	___
5.	Are illumination levels adequate in:				
	. Confined spaces?	___	___	___	___
	. Control rooms?	___	___	___	___
	. Laboratory?	___	___	___	___
	. Maintenance shops?	___	___	___	___
	. Manufacturing areas?	___	___	___	___
	. Offices?	___	___	___	___
	. Off-site locations?	___	___	___	___
	. Powerhouse?	___	___	___	___
	. Process areas?	___	___	___	___
	. Warehouse?	___	___	___	___
	. Other locations (________________)?	___	___	___	___
6.	Are lighting survey findings a factor in determining the frequency of lighting maintenance?	___	___	___	___

N/A - Not Applicable

WORKPLACE HEALTH PROTECTION PROGRAM ELEMENT CHECKLIST ON LIGHTING

		Yes	No	Don't Know	N/A
7.	Does the periodic lighting maintenance program include:				
	. Lamp replacement?	___	___	___	___
	. Fixture cleaning?	___	___	___	___
	. Cleaning of surroundings (wall/ceiling) to maintain reflectance?	___	___	___	___
8.	Are fixtures effectively distributed so that harsh shadows are minimized?	___	___	___	___
9.	Are incandescent lamps provided at operations where a stroboscopic effect may occur on rotating equipment if fluorescent lights were used?	___	___	___	___
10.	Is general illumination adequate where supplemental (direct) lighting is provided?	___	___	___	___
11.	Are incandescent lamps provided as backup to mercury vapor lamps to provide illumination at exits and stairways where start-up delays of mercury vapor lamps occur?	___	___	___	___
12.	Is an effort made to identify and eliminate sources of glare?	___	___	___	___
13.	Is ballast noise a consideration in lighting system design for offices or other quiet areas?	___	___	___	___
14.	Are PCB containing ballasts identified/labeled?	___	___	___	___
15.	Is there an established procedure for the handling and disposal of PCB containing ballasts?	___	___	___	___
16.	Are PCB containing ballasts properly disposed?	___	___	___	___

N/A - Not Applicable

WORKPLACE HEALTH PROTECTION PROGRAM ELEMENT CHECKLIST ON LIGHTING

		Yes	No	Don't Know	N/A
17.	Is adequate electric lighting provided in areas where daylighting is the principal light source (i.e., for periods when daylighting may be inadequate)?	___	___	___	___
18.	Are there special lighting considerations for areas/offices where VDT equipment is in operation?	___	___	___	___
19.	The facility's objective in carrying out lighting surveys is to:				
	. Determine adequacy of illumination levels?	___	___	___	___
	. Determine quality of lighting?	___	___	___	___
	. Determine presence of harsh shadows?	___	___	___	___
	. Identify glare sources?	___	___	___	___
	. Identify changes needed to improve lighting:				
	- Surface reflectance?	___	___	___	___
	- Relamping needs?	___	___	___	___
	- Fixture cleaning needs?	___	___	___	___
	- Alternate fixture type needs?	___	___	___	___
	- Fixture relocation needs?	___	___	___	___
	- Need for higher wattage lamps?	___	___	___	___
	- Need for supplementary (direct) lighting?	___	___	___	___
20.	Is emergency lighting periodically tested?	___	___	___	___
21.	Is emergency lighting adequate in:				
	. Control rooms?	___	___	___	___
	. Laboratory?	___	___	___	___
	. Active operating areas?	___	___	___	___
	. Manufacturing areas?	___	___	___	
	. Offices?	___	___	___	___
	. Boiler areas?	___	___	___	___
	. Substations (switchgear rooms)?	___	___	___	___
	. Parking areas?	___	___	___	___
	. Other locations (_______________)?				

N/A - Not Applicable

WORKPLACE HEALTH PROTECTION PROGRAM ELEMENT CHECKLIST ON LIGHTING

	Yes	No	Don't Know	N/A
22. Are fluorescent tubes disposed:				
. Properly to eliminate exposure to dust/mercury?	___	___	___	___
. Using a tube breaker unit?	___	___	___	___
23. Where applicable, is the tube breaker unit kept covered and stored in a remote, non-occupied area?	___	___	___	___

N/A - Not Applicable

WORKPLACE HEALTH PROTECTION PROGRAM ELEMENT
ON
MONITORING EQUIPMENT

WORKPLACE HEALTH PROTECTION PROGRAM ELEMENT ON MONITORING EQUIPMENT

OBJECTIVE

To provide information for consideration in the selection, maintenance, and calibration of equipment to be used in evaluating workplace health hazards; determining the need for, or the effectiveness of exposure controls; identifying sources of contaminant release; determining appropriate respiratory protective equipment for use at a task; and determining the adequacy of procedures/work practices.

BACKGROUND

Quantitative determinations of employee exposure to chemical, physical and biological agents, and ergonomic factors may be necessary to assess workplace health risks. This can be accomplished by employing a variety of direct reading instruments or sampling/analytical procedures. The method and/or equipment selected for a particular facility should be based on exposure potential; purpose for monitoring; sensitivity, accuracy, and precision required; types of operations; the potential for interferences; type of exposure assessment to be performed (full shift, short term, or peak exposure); area hazard classification; and acceptability by the user.

Knowledge of the materials present in a facility, the method and location of their use, as well as the frequency and duration of use are necessary for the selection of appropriate monitoring equipment for assessing exposure to health hazards. The potential for interferences to be present should not be overlooked. Manufacturer's literature should be reviewed to determine their possible effect on instrument response. It is also beneficial to talk to equipment owners to determine their experience with specific instruments.

Limitations have been observed in the field when monitoring equipment was in use. For example, the response of a number of hydrogen sulfide monitors, equipped with electrochemical sensors, is affected by elevated hydrogen gas levels (500 ppm and above). Some carbon dioxide monitors are significantly affected by low concentrations of sulfur dioxide gas (e.g., 1-2 ppm). The response of some carbon monoxide monitors equipped with electrochemical sensors becomes sluggish when used in a cold environment (e.g., less than 0° C). Operation of two-way radios can affect instruments that are not provided with electromagnetic interference (EMI) shielding.

Purposes for monitoring include health hazard assessment, exposure documentation, evaluation of engineering controls, identification of the need for exposure controls, selection of personal protective equipment, determining compliance with regulations, response to complaints, identifying sources of contamination, and others. Different types of monitoring equipment can be used to develop a database on which to draw conclusions relative to the purpose for sampling.

CONSIDERATIONS

The equipment inventory should be developed based on the exposure factors of concern, purpose for monitoring, the scope of program activities, and availability of resources to use and maintain the equipment. Space must be available for maintenance, storage, and calibration of the monitoring equipment and users must be trained in its use and the interpretation of results. Instrument manuals should be readily available for reference.

Batteries (dry cell type) should be removed from instruments that will not be used for some time (e.g., 3-4 weeks). Rechargeable battery operated equipment should be maintained in accordance with manufacturer's instructions.

Detector tube stocks should be adequate and within date. If possible, the tubes should be kept in a refrigerator. Detector tube pumps should be checked for leakage periodically and records of results maintained.

Equipment (electronic flow calibrators, burettes, rotometers, calibrant gases, means to make up calibrants, etc.) should be available to perform periodic calibrations. Records should be maintained of the instrument calibrations and response checks.

Instruments in the equipment inventory should be calibrated periodically by the manufacturer or other resource that can provide a certificate of calibration traceable to an accepted standard (e.g., National Institute for Standards and Technology - NIST). These calibrations should be performed on an annual basis. Included would be instruments such as sound level meters and calibrators, airflow measuring devices, electronic flow calibrators, ionizing radiation meters, direct reading toxic gas/vapor monitors, and others.

For some industrial hygiene functions, it is necessary to obtain and analyze samples employing methodologies defined by regulatory agencies. Effective documentation and recordkeeping procedures should be in place to support such monitoring programs. Appropriate sample data sheets and laboratory-type notebooks are useful in meeting these needs.

REFERENCES

"Air Sampling Instruments for Evaluation of Atmospheric Contaminants", 7th Edition (Cincinnati, OH: American Conference of Governmental Industrial Hygienists, 1985).

Pagnotto, L. D. and R. G. Keenan, "Sampling and Analysis of Gases and Vapors" (Cincinnati, OH: The Industrial Environment - Its Evaluation and Control, U. S. Department of Health, Education and Welfare, National Institute for Occupational Safety and Health, 1973), pp. 167-180.

Keenan, R. G., "Direct Reading Instruments for Determining Concentration" (Cincinnati, OH: The Industrial Environment - Its Evaluation and Control, U. S. Department of Health, Education and Welfare, National Institute for Occupational Safety and Health, 1973), pp. 181-196.

"Inspection for Accident Prevention in Refineries", , Publication 2002, First Edition, (Washington, D.C: American Petroleum Institute, 1984).

"NIOSH Manual of Analytical Methods", Third Edition (Cincinnati, OH: U.S. Department of Health, Education and Welfare, National Institute for Occupational Safety and Health, 1984) pp. 43-64.

"Fundamentals of Industrial Hygiene", 3rd Edition (Chicago, Il: National Safety Council, 1988), pp 417-453.

WORKPLACE HEALTH PROTECTION PROGRAM ELEMENT CHECKLIST ON MONITORING EQUIPMENT

	Yes	No	Don't Know	N/A
1. Is there an up-to-date inventory of the facility's monitoring equipment?	___	___	___	___
2. Is the available equipment appropriate for assessing potential exposures associated with operations?	___	___	___	___
3. Have personnel been provided adequate training on the use, calibration, and maintenance of available monitoring equipment?	___	___	___	___
4. Is calibration equipment available:				
. Appropriate plastic bags?	___	___	___	___
. Burettes?	___	___	___	___
. Calibrant gas/vapor?	___	___	___	___
. Dry gas meter?	___	___	___	___
. Electronic flow calibrator?	___	___	___	___
. Large glass bottle (e.g., 20 L)?	___	___	___	___
. Radiation check source?	___	___	___	___
. Rotometers?	___	___	___	___
. Wet test meter?	___	___	___	___
. Calibrators for specific instruments?	___	___	___	___
5. Are adequate records maintained of instrument calibrations?	___	___	___	___
6. Are instrument operating manuals readily available?	___	___	___	___
7. Are instruments sent to an outside resource for calibration on a periodic basis:				
. Ionizing radiation meters? (Frequency -)	___	___	___	___
. Sound level meters/octave band analyzers? (Frequency -)	___	___	___	___
. Sound level calibrators? (Frequency -)	___	___	___	___
. Noise dosimeters/calibrators? (Frequency -)	___	___	___	___
. Airflow measuring devices? (Frequency -)	___	___	___	___

N/A - Not Applicable

WORKPLACE HEALTH PROTECTION PROGRAM ELEMENT CHECKLIST ON MONITORING EQUIPMENT

	Yes	No	Don't Know	N/A
. Light measuring devices? (Frequency -)	___	___	___	___
. Microwave detectors? (Frequency -)	___	___	___	___
. Other? (Frequency -)	___	___	___	___
8. Is there adequate space for maintenance and storage of equipment?	___	___	___	___
9. Are monitoring instruments effectively maintained?	___	___	___	___
10. Is there a capability to make oxygen determinations before hydrocarbons are measured in areas where the O_2 level may be sufficiently low to affect the hydrocarbon monitor response?	___	___	___	___
11. Have batteries (dry cell type) been removed from infrequently used equipment?	___	___	___	___
12. Are rechargeable battery operated devices effectively handled to prevent their developing poor battery memory characteristics?	___	___	___	___
13. Are detector tubes in date?	___	___	___	___
14. Are detector tube pumps checked for leakage before use?	___	___	___	___
15. Are detector tube pumps volumetrically checked?	___	___	___	___
16. Is available monitoring equipment state of-the-art?	___	___	___	___

N/A - Not Applicable

WORKPLACE HEALTH PROTECTION PROGRAM ELEMENT
ON
MONITORING PROGRAM

WORKPLACE HEALTH PROTECTION PROGRAM ELEMENT ON MONITORING PROGRAM

OBJECTIVE

To provide information for developing and implementing an effective monitoring program for assessing employee exposure to workplace hazards and preventing adverse health effects.

BACKGROUND

An assumption in the field of occupational health is that there is a concentration or limit for each contaminant or physical agent below which exposure to it will not result in an adverse health effect. Such exposure limits have been developed for many of the substances and physical agents to which individuals may be exposed in the work environment. There is, however, recognition that, due to variation in individual susceptibility, a small percentage of those exposed to a hazardous agent at or below an established limit may be adversely affected.

An assessment of employee exposure risk requires a determination of the magnitude of exposure to the agent and comparison of the result to its established limit. This is generally accomplished through the development and implementation of a monitoring program which is directed toward protecting employee health through the quantitative assessment of employee exposures to hazardous agents and reducing exposures, where warranted.

There are a number of reasons why a monitoring program is implemented. These include:

- For risk assessment by comparing exposure levels to established limits
- To respond to reports of excessive exposures
- To respond to employee concerns and to inform employees of the results of their exposure to potential health hazards
- To determine the effectiveness of engineering and administrative controls
- To acquire data for use in epidemiology studies
- To identify sources of contamination, such as from waste operations, fugitive emissions, etc.
- To provide information to medical for use in surveillance programs
- To respond to medical requests relative to illness investigations
- To identify the need for respirator use and selection of the appropriate type
- To respond to regulatory requirements
- To establish new limits, when warranted.

Information needed for monitoring program development includes:

- Inventory of materials used or produced
- Knowledge of the potential hazards to which personnel may be exposed
- Knowledge of the process and exposure potential of personnel to the chemical, physical or biological agents and ergonomic factors associated with the tasks being performed
- A prioritization of those hazards to be evaluated, based on the toxicity of the contaminant of concern and a qualitative assessment of the exposure hazard
- Applicable exposure limits (8-hour time-weighted average, short-term, ceiling value)
- Appropriate sampling methods for evaluating the exposures
- Number of personnel to be monitored

Priority for developing and implementing a monitoring program may be based on:

- An overexposure occurrence
- Toxicity of the materials to which personnel are exposed
- Anticipated exposure level
- Number of personnel potentially exposed to the hazardous agent

When the foregoing information is available, a monitoring program strategy can be developed to assess exposures. Initially, this could involve the determination of who to sample, the type of samples to obtain, when and how to collect the samples, etc. In addition, a decision logic should be developed for stopping or continuing the sampling program. A simple decision matrix is presented as Figure 1.1 in the referenced NIOSH sampling strategy manual. It is useful when applied to specific substance regulations and is adaptable for use in monitoring programs for non-regulated contaminants. It focuses on the identification and exposure assessment of the maximum risk employee in each work area in which there is a potential for exposure to the health hazard. In addition, there is information presented in the aforementioned reference for determining the number of personnel to sample from a group in order to have confidence that the results are representative for the exposed group.

CONSIDERATIONS

An effective monitoring program requires the use of reliable, calibrated equipment; validated sampling and analytical methods; and the analysis of samples by a reliable laboratory. In addition, personnel to be monitored should be divided into homogeneous exposure groups (i.e., have the same opportunity to be exposed to the contaminant) and be selected randomly from such groups. Dates on which samples are to be collected should also be based on a random selection process.

Personnel performing the monitoring should develop a knowledge of the processes being carried out and be aware of the potential contaminants that may be associated with various operations. In addition, there should be an understanding of the exposure potential for the contaminants that may be associated with the multitude of jobs/tasks that personnel perform. This can be achieved by conducting a qualitative exposure assessment of the operation/job assignment/task.

A determination must be made regarding the type of monitoring method(s) to be employed to assess a health hazard. This involves deciding whether the collection of short-term or long-term samples, or both, are necessary to characterize the exposure risk. For some health hazards, particularly acutely-acting contaminants, short-term samples may be required for comparing employee exposure with a short-term exposure limit (STEL) or ceiling limit. Decisions must be made regarding when to sample; the appropriate sampling time; and whether personal, exposure zone (hearing, breathing, body core, etc.), or area samples are appropriate; and whether task or job assignment exposure assessments, or both, should be made to assess exposure risk. In addition to the foregoing, determinations must be made as to who to sample, as well as how many individuals in a group of exposed personnel should be sampled to obtain a reliable estimate of exposure risk.

The individual(s) performing the monitoring should be thoroughly familiar with the equipment to be used and document all necessary information associated with the samples being obtained. Contact should be made with the laboratory to ensure that the sampling method being used (collection device, sample rate, sample time, etc.) is effective for the conditions which exist (humidity, temperature, potential interferences, etc.) and that the collected sample can be handled properly. This will minimize invalid samples and the suspicion by monitored personnel that there was something to hide when there is no result to report.

A considerable amount of time can be involved in responding to regulatory compliance monitoring programs. The exposure assessment requirements are generally defined within the context of workplace health protection standards. These can often be met by identifying and monitoring the maximum risk employee in each work area. If the exposure of this individual is acceptable, then others would not have to be monitored. The basis for the selection of the maximum risk employee should be documented. Use of this approach can be applied to non-regulatory monitoring programs as well.

A decision matrix could be developed for determining when monitoring of additional personnel is needed, as well as when monitoring can be discontinued. This is often keyed to an action level that is specified in some regulatory workplace health protection standards. A similar approach could be applied to the results for non-regulated contaminants. For example, if exposures are less than one-half of the 8-hour time-weighted average exposure limit, the need for repeat monitoring would likely be unnecessary unless there was a change in the process or task(s). Other factors to be considered in determining repeat sampling needs and sampling frequency include the toxicity of the contaminant, exposure control effectiveness, number of personnel exposed to the hazard, variability of sample results, availability of a more reliable minitoring method, and others.

If the substance being monitored is one which is capable of being absorbed through the skin, as well as by inhalation, then attention should be given to skin contact as a route of exposure. The control measures to be implemented, when necessary, should be directed toward minimizing skin contact, as well as inhalation of the contaminant. If warranted, the employee could be referred to medical for biological monitoring to determine if the dose of contaminant received through skin absorption and inhalation was excessive. Biological Exposure Indices (BEIs) are available for a number of contaminants and the application of these may be useful as a determinant of the relative contribution of dermal exposure to the total dose received by the exposed individual. However, other factors could affect the result and these must be considered.

Sample results should be provided to management, supervisory personnel, and the monitored employee, as well as those who perform the same task(s). This should be done in a timely manner as agreed to by management, or as required by regulation. In addition, monitoring records should be properly maintained. If a computer system is used, the transfer of data into the system should be accurate and timely. Field sampling data sheets should not be sent to storage until the data entry has been reviewed, and found satisfactory for completeness and accuracy.

A quality assurance protocol could be incorporated into the monitoring program. It may include specifications for instrument calibration and flow checks (e.g., before, during, and after sample collection, as appropriate); documentation of data on standardized field data sheets; use of blank samples, as well as spiked and/or replicate samples, where practical; splitting of samples, where practical; requirements for storage and shipment of samples to the laboratory to ensure sample integrity; and a quality control program to ensure data is correct. Where appropriate, a chain of custody procedure should be incorporated into the monitoring program. In addition, the laboratory to which samples are submitted for analysis should have an effective quality assurance program.

A report of findings of the monitoring results, including recommendations for reducing exposures, when necessary, could be prepared and distributed per management's direction. A copy should be retained for reference and follow-up, as appropriate.

REFERENCES

"Lynch, J. R., "Measurement of Worker Exposure", Patty's Industrial Hygiene and Toxicology, Volume III, Chapter 6, Theory and Rationale of Industrial Hygiene Practice (New York: John Wiley & Sons, Inc., 1979).

"Occupational Exposure Sampling Strategy Manual" (Cincinnati, OH: U. S. Department of Health, Education and Welfare, National Institute for Occupational Safety and Health, 1977).

"Threshold Limit Values for Chemical Substances and Physical Agents and Biological Exposure Indices" (Cincinnati, OH: American Conference of Governmental Industrial Hygienists, Latest Edition).

"Fundamentals of Industrial Hygiene" (Chicago, IL: National Safety Council, 1979), pp. 465-611.

Soule, R. D., "Industrial Hygiene Sampling and Analysis", Vol. 1, General Principles, Patty's Industrial Hygiene and Toxicology (New York: John Wiley & Sons, Inc., 1989), pp. 707-769.

"Air Sampling Instruments", 7th Edition (Cincinnati, OH: U.S. Department of Health, Education and Welfare, National Institute for Occupational Safety and Health, 1989).

Rappaport, S. M., "Assessment of Long-Term Exposures to Toxic Substances in Air", Annals of Occupational Hygiene, 35(1): 61-121, 1991.

WORKPLACE HEALTH PROTECTION PROGRAM ELEMENT CHECKLIST ON MONITORING PROGRAM

		Yes	No	Don't Know	N/A
1.	Is there a complete inventory of materials used and produced?	___	___	___	___
2.	Are personnel who perform monitoring knowledgeable with respect to the potential health hazards that may be associated with each operation?	___	___	___	___
3.	Have qualitative exposure assessments been carried out for jobs/tasks that are to be included in the monitoring program?	___	___	___	___
4.	Are the objectives of the monitoring program developed before sampling sessions are initiated?	___	___	___	___
5.	Have decisions been made beforehand regarding who will be sampled and the types of samples to be collected?	___	___	___	___
6.	Is a determination made that validated sampling/analytical methods are available (when appropriate) for:				
	. 8-hour time-weighted average samples?	___	___	___	___
	. Short-term (e.g., 15 minute) samples?	___	___	___	___
	. Ceiling limit exposure assessments?	___	___	___	___
7.	Are homogeneous exposure groups identified for inclusion in the program?	___	___	___	___
8.	When appropriate, has the basis for identifying the highest risk employee been documented?	___	___	___	___
9.	Are personnel to be monitored randomly selected?	___	___	___	___
10.	Are dates on which sampling will be conducted randomly selected?	___	___	___	___

N/A - Not Applicable

WORKPLACE HEALTH PROTECTION PROGRAM ELEMENT CHECKLIST ON MONITORING PROGRAM

	Yes	No	Don't Know	N/A
11. Are samples obtained in each season of the year?	___	___	___	___
12. Is the sampling effort directed toward obtaining representative interday exposure profiles for homogeneous groups?	___	___	___	___
13. Is a quality assurance aspect included in the monitoring program?	___	___	___	___
14. Are blank samples submitted to the laboratory with each sample set?	___	___	___	___
15. When practical, are spiked samples periodically submitted to the laboratory with samples?	___	___	___	___
16. Are replicate samples obtained in each sample session as a quality assurance check on the sampling/analysis procedures?	___	___	___	___
17. Are samples submitted to an accredited laboratory?	___	___	___	___
18. Has the laboratory's quality assurance/ quality control program been reviewed?	___	___	___	___
19. Is there communication with the laboratory to determine sampling parameters for assessing exposure by the types of samples to be obtained (e.g., 8-hour TWA, STEL, etc.)?	___	___	___	___
20. Is the laboratory apprised of circumstances which could possibly invalidate a sampling/analytical method (e.g., humidity, temperature, interference, etc.)?	___	___	___	___
21. Is adequate, reliable sampling equipment available for carrying out a monitoring program?	___	___	___	___

N/A - Not Applicable

WORKPLACE HEALTH PROTECTION PROGRAM ELEMENT CHECKLIST ON MONITORING PROGRAM

		Yes	No	Don't Know	N/A
22.	Is monitoring equipment calibrated before and after use?	___	___	___	___
23.	Are instrument calibration data adequate and documented?	___	___	___	___
24.	Are individuals who carry out the monitoring program knowledgeable regarding equipment use?	___	___	___	___
25.	Have measures been taken to minimize the occurrence of "lost samples"?	___	___	___	___
26.	Is there an established procedure for determining the need for continued monitoring or cessation of sampling?	___	___	___	___
27.	Are recommendations made to medical for biological testing of an individual based on a monitoring result?	___	___	___	___
28.	Are other routes of exposure (e.g., skin absorption) considered during a personal monitoring study?	___	___	___	___
29.	Have any employees been recommended (if appropriate for a specific contaminant) to medical for biological monitoring to determine the dermal exposure contribution to the total dose received?	___	___	___	___
30.	Are monitoring results provided to those employees whose exposures have been evaluated?	___	___	___	___
31.	Are monitoring results of representative employees provided to other personnel who have the same job assignment?	___	___	___	___
32.	Are standardized forms used in the monitoring program for data collection?	___	___	___	___

N/A - Not Applicable

WORKPLACE HEALTH PROTECTION PROGRAM ELEMENT CHECKLIST ON MONITORING PROGRAM

		Yes	No	Don't Know	N/A
33.	Are monitoring data sheets properly completed?	___	___	___	___
34.	Are monitoring data accurately entered into a recordkeeping system in a timely manner?	___	___	___	___
35.	Are monitoring data reviewed by a qualified individual?	___	___	___	___
36.	Are investigations carried out when an exposure limit (TWA, STEL, Ceiling value) is exceeded?	___	___	___	___
37.	Are hard copies of monitoring data sheets effectively retained for reference? (Where?__________________)	___	___	___	___
38.	Are monitoring records and results sufficiently organized for use in:				
	. Assessing the impact of a future regulation?	___	___	___	___
	. Discussing findings with supervisor/employee?	___	___	___	___
	. Epidemiology studies?	___	___	___	___
	. Determining the potential impact of a reduction in an exposure limit?	___	___	___	___
	. Comparing monitoring results with those of other similar facilities?	___	___	___	___
	. Statistical analyses?	___	___	___	___
	. Legal issues?	___	___	___	___
39.	Are reports of monitoring results provided to management?	___	___	___	___
40.	Are reports of monitoring results readily available?	___	___	___	___
41.	Is there a practice to conduct follow-up monitoring to ensure that control measures that have been implemented are effective?	___	___	___	___

N/A - Not Applicable

WORKPLACE HEALTH PROTECTION PROGRAM ELEMENT CHECKLIST ON MONITORING PROGRAM

	Yes	No	Don't Know	N/A
42. Are there continuous monitors in operation within the facility?	___	___	___	___
43. Are records (i.e., response checks, calibrations, alarm situations) of the operation of continuous monitors effectively maintained?	___	___	___	___

N/A - Not Applicable

WORKPLACE HEALTH PROTECTION PROGRAM ELEMENT
ON
NON-IONIZING RADIATION

WORKPLACE HEALTH PROTECTION PROGRAM ELEMENT ON NON-IONIZING RADIATION

OBJECTIVE

To provide general information for the development and implementation of a program to assess non-ionizing radiation exposure potential and identify control needs to reduce exposures, where appropriate.

BACKGROUND

The range of wavelengths of the electromagnetic spectrum extends from less than 10^{-12} centimeters (cm) to greater than 10^{10} cm. Cosmic rays and X-radiations have the shortest wavelength and are ionizing radiations. Ultraviolet (UV), visible, infrared (IR), microwaves and the radio frequencies have longer wavelengths and make up the non-ionizing radiation sources to which personnel may be exposed.

The photon energies of the radiations in the electromagnetic spectrum are directly proportional to their frequency and inversely proportional to their wavelength. Thus, the higher energies are associated with X-rays and the lower energies with radio frequency waves. The photon energy considered necessary to ionize matter, such as atomic oxygen or hydrogen, is about 12 electron volts (eV). Electromagnetic radiation sources with photon energies less than 12 eV are not considered capable of producing ionization in matter and do not present a ionizing radiation exposure potential to personnel. Thus, they are referred to as non-ionizing radiations.

The biological effects resulting from exposure to non-ionizing radiation depends upon the wavelength and radiant energy of the source, dose rate, individual susceptibility, and the total energy (dose) incident on the exposed area within a given time interval.

The eye is considered the organ at greatest risk of damage from exposure to non-ionizing radiation. Skin effects are also of significant concern for exposures to several types of non-ionizing radiation. Other adverse health effects, alleged to be the result of exposure to the electric or magnetic fields associated with electromagnetic radiation sources, are presently under investigation.

Ultraviolet Radiation

Ultraviolet radiation occupies that region of the electromagnetic spectrum between X-rays and visible light. The range of interest, from a workplace health standpoint, extends from 100 nanometers (nm) to 400 nm. The ultraviolet region of the electromagnetic spectrum has been divided into the UV-A or near-region (315-400 nm), the UV-B or mid-region (280-315 nm), and the UV-C or far-region (100-280 nm). The UV-C region is referred to as the germicidal region. The UV-B region presents the greatest hazard potential for producing adverse health effects.

Typical sources of exposure to UV radiation include the sun, welding operations and others. Potential adverse health effects associated with exposure to these include skin reddening (erythema) and the sensation of sand in the eyes (photokeratitis). These effects are primarily associated with exposure to UV energy at wavelengths between 200 nm and 310 nm (i.e., the UV-B region, primarily, with some effects from photons at the upper end of the UV-C region).

Skin cancer has occurred as a result of outdoor work in the sun. This is considered a UV radiation exposure effect. Concurrent exposure to some types of chemicals (coal tar derivatives, polynuclear aromatic compounds, etc.) is believed to increase the potential for skin cancer to develop among personnel working in the sun. In addition, photosensitization may result when there is concurrent exposure to these type chemicals and UV radiation.

Ultraviolet radiations are absorbed by a wide variety of materials. Thus, adverse eye and skin effects can be prevented through the use of appropriate eye protection and clothing to prevent exposure to the UV radiations. Window glass is opaque to UV radiations. Commercially available sunscreens can be applied to the skin to absorb UV radiations when there is skin exposure to the sun. Screens can be employed at welding operations to prevent exposures of non-operating personnel to the radiations associated with this work.

Exposure limits have been developed by the American Conference of Governmental Industrial Hygienists for ultraviolet radiations in the wavelength range of 180 to 300 nm. They apply to eye and skin exposures from most UV sources except lasers, photosensitive individuals, and where there is concomitant exposure to photosensitizing agents.

Infrared Radiation

Infrared (IR) radiation occupies that portion of the electromagnetic spectrum between visible light and the radio frequency region. It has been divided into the IR-A or near-infrared region (760-1,400 nm), the IR-B or intermediate region (1,400 to 3,000 nm), and the IR-C or far-infrared region (3,000 to 10^6 nm). The range of interest from a workplace health concern standpoint, extends from 760 nm to about 1,500 nm. Infrared radiation is emitted by a number of sources including the sun, operating furnaces, welding arcs, and hot surfaces. The emission characteristics (wavelength and energy) are largely dependent on source temperature. The mechanism for IR radiations to produce a biological effect involves the absorption of the IR radiation and the consequent heating of tissue. The body organs at greatest risk from IR radiation exposure are the skin and eyes.

IR radiation is perceptible as a sensation of warmth. It causes increased tissue temperature. Thus, the principal effect of infrared radiation on the skin and eye is thermal injury. Infrared radiations in the near-IR range can damage the retina of the eye. The iris may be damaged by IR radiations below 1,000 nm and thermal damage to the cornea can result from absorption of energy in the epithelium. Exposure to infrared radiations reportedly can result in cataract formation.

Skin effects from exposure to infrared radiations include vasodilation, increased pigmentation, and tissue damage. The latter is dependent on skin temperature and duration of exposure. The threshold of pain for skin is reached at a skin temperature of 114°F (45°C).

Infrared radiations are absorbed by a variety of materials. Thus, shielding can be used to reduce exposures to IR radiations. Highly reflective (low absorptivity) shields are the most effective type. Reducing source emissivity and the use of reflective clothing can also be employed to reduce personnel exposures to IR sources.

Exposure limits have been developed for the visible light and near-infrared radiation regions of the electromagnetic spectrum (the wavelength range of 400 to 1400 nm). They were developed to protect against eye effects and are discussed in the booklet "Threshold Limit Values for Chemical Substances and Physical Agents and Biological Exposure Indices", published by the American Conference of Governmental Industrial Hygienists.

Radio Frequencies

The radio frequency (RF) range of the electromagnetic spectrum extends from about 100 kilohertz (kHz) to 300 gigahertz (GHz). It includes the frequencies at which radar, microwave, and diathermy equipment is operated, as well as those at which television and radio transmissions are made. The RF range relative to the potential to produce adverse health effects is from 10 megahertz (MHz) to 300 GHz. The American National Standards Institute has published Radio Frequency Protection Guides (RFPGs) to eliminate adverse effects on the functioning of the human body as a result of exposure to RF energy. These apply to the associated electric and magnetic field strengths and power density, as a function of frequency.

RF energy absorption by biological systems is a function of frequency, orientation, and the dielectric properties of the absorber. The primary modes of action of these non-ionizing radiations on a biological system include electron excitation, molecular dissociation, formation of free radicals, and production of heat.

The frequency range of microwaves is from 30 MHz to 300 GHz. Depending on frequency, microwave energy can be reflected by the body (i.e., the higher frequencies), absorbed at varying depths in body tissue (i.e., the mid-frequencies), or pass directly through the body without effect (i.e., low frequency microwaves).

When absorbed, microwave radiations induce localized heating. Skin absorption of microwave energy can produce pain, thereby warning one of the exposure. However, if the microwave energy is absorbed in deeper organs, there may be a rise in tissue temperature with little awareness of the heat being generated or of the consequent injury. Effects are dependent on the microwave frequency/wavelength; power density; length of exposure; geometry, size and orientation of the irradiated object; and the nature of the tissue being irradiated. Shorter wavelength microwaves are reportedly absorbed at the skin surface and longer waves in deeper tissue. The most susceptible body organs are those which have a high blood supply such as muscle, the brain, and skin. Cataract formation has been noted in animal studies where exposure to a specific frequency range (i.e., 1-10 GHz) occurred.

Exposure limits have been established by the American Conference of Governmental Industrial Hygienists for radio frequency radiation, including the microwave region of the spectrum. These specify limits for associated electric and magnetic field strengths, as well as power density limits for specific frequencies of RF radiation.

Electromagnetic radiations in the radiofrequency range can affect electronic devices including older type heart pacemakers, a number of types of industrial hygiene monitoring devices, computer controls, and others. For example, portable radio transmitters (in the transmit mode), such as those used by personnel in petroleum and petrochemical process operations, have been demonstrated to affect some types of personal and portable industrial hygiene monitoring equipment.

Lasers

The Laser Institute of America defines the laser as an intense, highly directional beam of light that, if directed, reflected, or focused upon an object, will be partially absorbed, thereby raising the temperature of the surface and/or the interior of the object and potentially causing an alteration or deformation of the material. Others refer to a laser as a device that is used to generate and emit coherent electromagnetic radiation at wavelengths in the optical region of the electromagnetic spectrum.

The term "laser" is an acronym for "light amplification by stimulated emission of radiation". The lasing medium may be a gas, crystalline material, a liquid, or a semiconductor material. The wavelength of the light beam that is emitted, which is based on the lasing media, may be in the ultraviolet, visible, or infrared region. Laser units are operated in a pulsed, continuous wave (CW), or Q-switched mode depending on their design. The operating mode determines, to some degree, the power of the beam emitted by the laser unit.

The biologic effects of lasers depends upon their wavelength, source size, beam divergence, radiant exposure, duration of exposure, environmental conditions, and susceptibility of those exposed. The eye and the skin are the critical organs for laser radiation exposure. The human eye is almost always more vulnerable to injury than human skin. Thus, if laser radiation is controlled to levels at which adverse effects to the eye will not occur, then the hazard to the skin will also be controlled. Little is known concerning the chronic effects of laser exposure. Exposure standards have emphasized the need for protecting against acute eye exposures.

The eye is most vulnerable to laser radiations in the visible and infrared regions. Visible and IR-A (near IR) laser wavelengths deposit their energy at the back of the eye (e.g., the retina), whereas other laser wavelengths (UV, IR-B/C) deposit their energy in the cornea, conjunctiva and other anterior structures of the eye producing photokeratitis, conjunctivitis or thermal damage to the lens. Skin effects may include erythema or blistering/charring depending upon laser power, wavelength, exposure time, etc. The UV and IR-B/C laser beam wavelengths are those which produce these skin effects.

Exposure limits have been developed for eye and skin exposure to laser beams. These are presented in the Threshold Limit Values for Chronic Substances and Physical Agents and Biological Exposure Indices booklet, published by the American Conference of Governmental Industrial Hygienists.

A practical means for evaluating and controlling laser radiation exposure is to classify laser systems according to their relative hazards, specify the controls appropriate for each classification (e.g., Class 1,2,3, etc.), and assess the potential for the generation of other hazards when the laser is operational. Exposure control recommendations for various laser classes are presented in Table 10 of the ANSI Z136.1 (1986) standard for the "Safe Use of Lasers". Since environmental and personnel factors can play a role in determining the laser exposure hazard, additional measures may need to be implemented. These could involve the identification of a laser safety officer and a review of all laser operations before they are energized to ensure that exposure controls are effective.

For some laser systems, operating personnel should be trained. In addition, access to some operating areas should be restricted, hazard warning signs posted, interlocks installed on laser system access doors, the beam path enclosed, beam stops installed, reflecting surfaces eliminated in the operating area and personnel provided, and required to wear, appropriate eye protection. Required eyewear characteristics are to be based on the wavelength of the laser beam and the laser beam irradiance.

CONSIDERATIONS

Personnel potentially exposed to non-ionizing radiation should be made aware of the hazards and sources of potential exposure. Equipment that may be a source of hazardous exposure to non-ionizing electromagnetic energy should be designed, installed, operated, and maintained in a manner which reduces emissions and exposures to acceptable levels. Engineering controls should be properly maintained to ensure their effectiveness in reducing emissions and consequent exposure. Established procedures/work practices should be adhered to in operations, and protective equipment provided personnel, when appropriate.

The hazard potential associated with exposure to each source of non-ionizing radiation should be evaluated and exposure controls implemented, where appropriate. This may involve a review of the work task, determination of exposure time and frequency of exposure, assessment of the effectiveness of engineering and administrative controls, and the use of appropriate personal protective equipment. Where necessary, quantitative assessments should be carried out to determine exposure risk.

Welding operations (e.g., electric arc, plasma arc, inert gas, etc.) can be a source of exposure to ultraviolet and infrared radiation, and intense visible light. Eye and skin protection should be worn when welding to reduce the potential for adverse effects to occur. Welding screens, or other type barrier, can be positioned around welding areas to prevent the exposure of non-welding personnel to the non-ionizing radiation hazards associated with the operation.

Personnel may be exposed to infrared radiation and intense visible light while inspecting or replacing furnace burners, checking furnace tubes, doing oxyacetylene burning, and at other operations where exposure to flames or extremely hot surfaces may occur. Appropriate eye and skin protection should be provided personnel to reduce the associated exposure hazard. Infrared radiation exposures also occur when personnel are exposed to radiant heat sources such as heat exchangers, steam driven equipment (pumps, compressors, etc.), hot lines, entering furnaces for inspection/work, cleaning furnace tubes, etc. Measures need to be taken to reduce the potential for excessive heat stress effects to develop among personnel performing these tasks. This may involve providing ventilation, installing shielding, insulating hot equipment, etc.

Exposure to electromagnetic radiation in the microwave range of the spectrum may occur in association with microwave oven operation, use of diathermy equipment, during microwave communication system transmissions and radar operations, as well as in the maintenance of this equipment. Where appropriate, signs should be used to alert personnel of the potential RF hazard. Leakage from microwave ovens should be evaluated periodically and interlocks checked to ensure their effectiveness. Leakage radiation from older types of diathermy equipment reportedly can be significant and excessive exposures to microwave energies may result from their operation. Leakage suppression from diathermy equipment can be achieved by replacement of the applicator and shielding the cables.

The RF radiation hazard potential associated with exposure to electromagnetic radiations from antennas depends upon several factors including the average radiated power emitted by the antenna; size, shape and type of the antenna; distance from the antenna; and wavelength. Formulas, based on distance from the antenna, can be employed to compute the expected power density for estimating the exposure risk. However, other factors, such as standing waves, reflections, presence of multiple sources, etc. can affect the exposure risk.

A laser safety program should be developed if lasers are to be operated in a facility. If Class 3 or Class 4 lasers are to be used it would be appropriate to identify a laser safety officer (LSO) and require that written procedures be developed for the operation of these devices. The LSO would have responsibility to review proposals for the use of lasers and approve the written procedures to be followed. This individual should also review operations to ensure adherence with written procedures and also ensure that the associated exposure potential is low.

All laser operations should be reviewed to assure that exposure hazards are effectively controlled through the application of engineering and administrative controls. These include enclosures; interlocks; use of warning lights; limiting access to the operating area; locating the laser beam at an acceptable height (where eye contact is unlikely); assuring proper laser alignment; eliminating reflective surfaces in the work area; providing beam stops; posting signs; and ensuring personnel are adequately trained to operate the equipment safely. When appropriate, personal protective equipment should be provided. The eye protection made available to personnel should be carefully selected, based on the characteristics of the laser, and it should be inspected periodically to ensure its continued effectiveness. When warranted, eye examinations should be provided to laser operating personnel.

REFERENCES

"Radio Frequency Hazard Warning Symbol", ANSI C95.2-1982 (New York: American National Standards Institute, 1982).

"Safety Levels with Respect to Human Exposure to Radio Frequency Electromagnetic Fields, 300 kHz to 100 GHz", ANSI C95.1-1982 (New York: American National Standards Institute, 1982).

"Techniques and Instrumentation for the Measurement of Potentially Hazardous Electromagnetic Radiation at Microwave Frequencies", ANSI C95.3-1973 (New York: American National Standards Institute, 1973).

"Recommended Practice for the Measurement of Hazardous Electromagnetic Fields - RF and Microwave", ANSI C95.5- 1981 (Washington, D. C.: American National Standards Institute, 1981).

"Performance Standards for Microwave and Radio Frequency Emitting Products", 21 CFR Part 1030 (Washington, D. C.: U.S. Food and Drug Administration, April 1990).

"Threshold Limit Values for Chemical Substances and Physical Agents and Biological Exposure Indices", Latest Edition (Cincinnati, OH: American Conference of Governmental Industrial Hygienists).

Wilkening, G. M., "Nonionizing Radiation", Patty's Industrial Hygiene and Toxicology - General Principles (New York: John Wiley & Sons, 1978), pp. 359-440.

"Safe Use of Lasers", ANSI Z135.1-1986 (New York: American National Standards Institute, 1986).

"Laser Safety Guide" (Orlando, FL: Laser Institute of America, 1989).

"Performance Standard for Light-Emitting Products" (Washington, D. C.: U.S. Food and Drug Administration, April 1990).

Bassen, H. I., "Radio-Frequency and Microwave Radiation", National Safety News, October 1980, pp. 57-64.

Sliney, D. and H. LeBodo, "Laser Eye Protectors", Journal of Laser Applications, Summer/Fall 1990, pp. 9-13.

Elza, D., "You Can Live With Lasers", National Safety and Health News, May 1985, pp. 49-51.

"A Guide to the Control of Laser Hazards" (Cincinnati, OH: American Conference of Governmental Industrial Hygienists, 1990).

Polk, C. and E. Postbow, Ed., "Handbook of Biological Effects of Electromagnetic Fields" (Boca Raton, FL: CRC Press, Inc., 1986).

"Guidelines for Laser Safety and Hazard Assessment", OSHA Instruction PUB 8-1.7 (Washington, D. C.: United States Dept. of Labor, 1991).

WORKPLACE HEALTH PROTECTION PROGRAM ELEMENT CHECKLIST ON NON-IONIZING RADIATION

	Yes	No	Don't Know	N/A
1. Have non-ionizing radiation sources been identified:				
. Communication systems?	___	___	___	___
. Flames/hot surfaces?	___	___	___	___
. Diathermy units?	___	___	___	___
. Lasers?	___	___	___	___
. Microwave ovens?	___	___	___	___
. Radar Equipment?	___	___	___	___
. Welding operations?	___	___	___	___
. Other?(____________________)	___	___	___	___
2. Have personnel been advised of the hazards associated with the electromagnetic radiations to which they may be exposed?	___	___	___	___
3. Are personnel who install, operate, adjust or maintain non-ionizing radiation generating equipment trained and qualified to perform the work?	___	___	___	___
4. Are procedures/work practices/protective equipment adequate to minimize exposure to non-ionizing radiation from:				
. IR radiation sources?	___	___	___	___
. Laser equipment?	___	___	___	___
. Microwave radiation sources?	___	___	___	___
. Radar equipment?	___	___	___	___
. UV radiation sources?	___	___	___	___
. RF radiation sources?	___	___	___	___
5. Are engineering controls adequate to limit exposure to non-ionizing radiation in the operation of:				
. Lasers?	___	___	___	___
. Microwave radiation sources?	___	___	___	___
. Diathermy units?	___	___	___	___
. RF sealers?	___	___	___	___
. Radar equipment?	___	___	___	___
6. Are microwave ovens periodically checked for:				
. Leakage?	___	___	___	___
. Interlock operation?	___	___	___	___

N/A - Not Applicable

WORKPLACE HEALTH PROTECTION PROGRAM ELEMENT CHECKLIST ON NON-IONIZING RADIATION

	Yes	No	Don't Know	N/A
7. Have determinations been made to assess the status of compliance with regulations/standards applicable to non-ionizing radiation sources?	___	___	___	___
8. Has a determination been made of the need to evaluate personnel exposure to electromagnetic radiation associated with the operation of non-ionizing radiation sources?	___	___	___	___
9. Is available monitoring equipment adequate to assess exposure to non-ionizing radiation sources in the facility?	___	___	___	___
10. Is appropriate protective equipment provided to personnel potentially exposed to:				
. IR radiation?	___	___	___	___
. Microwave radiation?	___	___	___	___
. Radio frequency radiation?	___	___	___	___
. UV radiation?	___	___	___	___
11. When appropriate, have measurements been made to determine the electric and magnetic field strengths associated with:				
. IR radiation sources?	___	___	___	___
. Laser equipment?	___	___	___	___
. Microwave radiation sources?	___	___	___	___
. Radar radiation sources?	___	___	___	___
. RF radiation sources?	___	___	___	___
. UV radiation sources?	___	___	___	___
12. Is laser use/application/facility design reviewed before a laser is purchased and energized?	___	___	___	___
13. Are lasers properly identified:				
. Class?	___	___	___	___
. Labels?	___	___	___	___

N/A - Not Applicable

WORKPLACE HEALTH PROTECTION PROGRAM ELEMENT CHECKLIST ON NON-IONIZING RADIATION

		Yes	No	Don't Know	N/A
14.	Have laser operators been provided training appropriate for safely operating equipment?	___	___	___	___
15.	Is there a written laser safety program?	___	___	___	___
16.	Is there a designated laser safety officer?	___	___	___	___
17.	Is the laser safety officer informed before the purchase of a new laser?	___	___	___	___
18.	Are the following controls in place for laser operations (as applicable):				
	. Access restrictions?	___	___	___	___
	. Eye protection requirements?	___	___	___	___
	. Barriers?	___	___	___	___
	. Shrouds/enclosures?	___	___	___	___
	. Beam stops?	___	___	___	___
	. Signs/labels/warning lights?	___	___	___	___
	. Master switch/key for Class 3/4 laser operation?	___	___	___	___
19.	Is the area in which lasers are operated satisfactory with respect to:				
	. Posting of signs?	___	___	___	___
	. Lighting (i.e., a well lighted area)?	___	___	___	___
	. Elimination of reflective surfaces?	___	___	___	___
	. Restricting access?	___	___	___	___
	. Beam height relative to eye level?	___	___	___	___
	. Warning signals?	___	___	___	___
20.	Is laser alignment effectively maintained?	___	___	___	___

N/A - Not Applicable

WORKPLACE HEALTH PROTECTION PROGRAM ELEMENT CHECKLIST ON NON-IONIZING RADIATION

		Yes	No	Don't Know	N/A
21.	Are workplace health concerns, other than that of the laser beam, assessed when reviewing future and ongoing laser operations?	___	___	___	___
22.	Are laser operating personnel instructed to remove reflective items (e.g., jewelry) when operating equipment?	___	___	___	___
23.	Are all lasers provided with appropriate warning labels?	___	___	___	___
24.	Where appropriate, is the proper type eye protection and protective clothing available and used by laser equipment operators?	___	___	___	___
25.	Is laser eye protection inspected periodically to ensure integrity?	___	___	___	___
26.	Are there written operating procedures for Class 3 and 4 laser use?	___	___	___	___
27.	Are Class 3 and 4 lasers operated and maintained by authorized personnel only?	___	___	___	___
28.	Are eye examinations provided laser operating personnel?	___	___	___	___

N/A - Not Applicable

WORKPLACE HEALTH PROTECTION PROGRAM ELEMENT
ON
SAMPLE ANALYSIS

WORKPLACE HEALTH PROTECTION PROGRAM ELEMENT ON SAMPLE ANALYSIS

OBJECTIVE

To provide information for selecting a laboratory to analyze industrial hygiene type samples and ensuring the quality of the data.

BACKGROUND

Samples obtained to evaluate employee exposure to hazardous substances are routinely submitted to in-house or commercial laboratories for analysis. It is essential that these be collected and analyzed in a consistent manner, employing procedures which have been demonstrated to be acceptable. That is, the methods used should have been validated through established testing procedures. In addition, the laboratory should demonstrate, on an on-going basis, its ability to do quality work when using these methods.

There are testing and accreditation programs in effect to which a laboratory should subscribe and participate in routinely. One such accreditation program is that administered by the American Industrial Hygiene Association (AIHA). If at all possible, the laboratory selected to perform analysis of samples should be a participant in a proficiency testing program administered by a recognized accrediting organization.

Mobil's Technical Services Industrial Hygiene Testing Laboratory is accredited. It meets all requirements for the analysis of specific types of samples for evaluating occupational exposures to airborne contaminants.

CONSIDERATIONS

The laboratory selected for analysis of samples that have been obtained for determining employee exposure to hazardous substances should be accredited by a recognized authority and have an effective quality assurance (QA) program in place. The QA program should include details on the policies of the laboratory; the analytical methods to be employed; the procedures/specifications to be adhered to in determining the quality of the work being performed; and a quality control effort that involves the daily monitoring of work and work product to ensure that analytical standards and specifications are being met. If practical, the laboratory should participate in round-robin testing programs to demonstrate proficiency in specific areas.

A copy of the laboratory's quality assurance/quality control (QA/QC) program (or a summary) should be obtained and reviewed to determine if it is satisfactory for facility needs. If appropriate, a visit could be made to the analytical laboratory to review the QA/QC program.

After analytical results are obtained, it is sometimes of interest to discuss the findings with laboratory personnel. Access should be possible to the laboratory director, supervisor, and the chemist who did the analysis, so that detailed information can be obtained, when necessary. The laboratory should have a sample tracking capability along with means to retain hard copy (when generated) of analytical data (e.g., chromatograms). This information should be accessible for at least 3 years.

If spiked samples are to be submitted to the laboratory as a part of the quality assurance program, arrangements should be made with the laboratory so that these can be prepared by a qualified individual. The spiked samples, along with blank samples, should be submitted to the laboratory with the field samples and the analytical results of the spike and blank samples used to determine laboratory performance. Field personnel should work closely with the laboratory in implementing this practice.

Periodic collection of duplicate samples should be considered. Results from these can be used as a check on the reproducibility of the sampling and analytical method and/or verification of a result.

REFERENCES

Juran, J. M., "Quality Control Handbook", Edition 2 (New York: McGraw Hill, 1962).

"Quality Assurance Manual for Industrial Hygiene Chemistry" (Akron, OH: American Industrial Hygiene Association, 1988).

"The Industrial Environment - Its Evaluation and Control" (Cincinnati, OH: U.S. Department of Health, Education, and Welfare, National Institute for Occupational Safety and Health, 1973), pp. 227-297.

Finley, K. J., "The Business of Labs", Asbestos Issues, (August 1990), pp. 22-26.

Kelley, W. D., "Quality Control", Patty's Industrial Hygiene and Toxicology, Volume 1, 3rd Edition (New York: John Wiley & Sons, Inc., 1978), pp. 1223-1261.

Clarke, A. N., et al., "The First Step Toward Quality Assurance", American Environmental Laboratory, (October 1990), pp. 9-14.

De Roos, F. L., et al, "Basic Ingredients For Assuring Quality in the Analytical Laboratory", American Environmental Laboratory (October 1990), pp. 16-24.

"Quality Assurance for Environmental Measurements" ASTM Special Technical Publication 876 (Philadelphia, PA: American Society for Testing and Materials, 1985).

WORKPLACE HEALTH PROTECTION PROGRAM ELEMENT CHECKLIST ON SAMPLE ANALYSIS

		Yes	No	Don't Know	N/A
1.	Is the laboratory that is performing the analyses of industrial hygiene samples accredited?	___	___	___	___
2.	Does the laboratory participate in a proficiency testing program?	___	___	___	___
3.	Is the laboratory proficient in the categories of interest to your sampling program (e.g., solvents, metals, etc.)?	___	___	___	___
4.	Are laboratory audits conducted periodically by an:				
	. Internal group?	___	___	___	___
	. External group?	___	___	___	___
5.	Are laboratory audit reports available for review?	___	___	___	___
6.	Do laboratory personnel meet the following qualifications:				
	. Laboratory Director - B.S. degree plus 5 years industrial hygiene laboratory experience?	___	___	___	___
	. Laboratory Supervisor - B.S. degree plus 5 years laboratory experience?	___	___	___	___
	. Laboratory Analyst - qualified by training?	___	___	___	___
7.	Does the laboratory have a written QA/QC program?	___	___	___	___
8.	Is a copy of the laboratory's QA/QC written program available for review?	___	___	___	___
9.	Is the laboratory's QA/QC program reviewed and updated annually?	___	___	___	___
10.	Do you have a copy of the laboratory's QA/QC program manual?	___	___	___	___

N/A - Not Applicable

WORKPLACE HEALTH PROTECTION PROGRAM ELEMENT CHECKLIST ON SAMPLE ANALYSIS

		Yes	No	Don't Know	N/A
11.	Would the laboratory permit a visit to review their operations and its QA/QC program?	___	___	___	___
12.	Are analytical procedures that are followed those which have been developed and validated by NIOSH or other qualified group?	___	___	___	___
13.	Is there an analytical procedure manual that is accessible to all the laboratory staff?	___	___	___	___
14.	Are changes to sampling/analytical procedures written into the laboratory's standard operating procedure manual?	___	___	___	___
15.	Are procedural changes to the manual reviewed and signed by the supervisor?	___	___	___	___
16.	Are copies of internally developed sampling/analytical procedures available for review?	___	___	___	___
17.	Are calibration curves checked frequently in routine analysis work?	___	___	___	___
18.	Are quality control charts developed to determine analytical procedure performance for:				
	. Individuals?	___	___	___	___
	. Group?	___	___	___	___
19.	Is a chain of custody procedure adhered to in submitting samples to the laboratory?	___	___	___	___
20.	Are spiked samples available from the laboratory to submit as blind samples along with regular samples?	___	___	___	___
21.	Are spiked samples periodically submitted blind to the laboratory, along with regular samples to determine laboratory performance?	___	___	___	___

N/A - Not Applicable

WORKPLACE HEALTH PROTECTION PROGRAM ELEMENT CHECKLIST
ON
SAMPLE ANALYSIS

		Yes	No	Don't Know	N/A
22.	Is it possible to obtain hard copy of analytical data (e.g., chromatogram, spectrograph data, etc.) within 3 years following analysis of a sample?	___	___	___	___
23.	Is it possible to discuss analytical results with:				
	. Laboratory director?	___	___	___	___
	. Laboratory supervisor?	___	___	___	___
	. Chemist?	___	___	___	___
24.	Are the following types of samples employed in analytical procedures:				
	. Internal standards?	___	___	___	___
	. Blanks?	___	___	___	___
	. Replicates?	___	___	___	___
	. Spikes:				
	- Laboratory?	___	___	___	___
	- Field?	___	___	___	___
25.	Are blank samples routinely submitted to the laboratory along with field samples?	___	___	___	___
26.	Are laboratory personnel adequately trained to perform analyses?	___	___	___	___
27.	Are samples sometimes taken in duplicate in order to obtain a second analysis for verification of a result?	___	___	___	___
28.	Are analytical balances calibrated yearly?	___	___	___	___
29.	Are sample results reviewed and validated before issuing a report of findings?	___	___	___	___
30.	Is sample turnaround time satisfactory for:				
	. Routine samples (e.g., benzene, asbestos, etc.)?	___	___	___	___
	. Non-routine samples (e.g., lead, nickel, etc.)?	___	___	___	___

N/A - Not Applicable

WORKPLACE HEALTH PROTECTION PROGRAM ELEMENT CHECKLIST ON SAMPLE ANALYSIS

	Yes	No	Don't Know	N/A
31. Is there a QA/QC program for samples which are analyzed in the facility (e.g., analyzing asbestos samples by field personnel at the plant)?	___	___	___	___

N/A - Not Applicable

WORKPLACE HEALTH PROTECTION PROGRAM ELEMENT
ON
VENTILATION

WORKPLACE HEALTH PROTECTION PROGRAM ELEMENT ON VENTILATION

OBJECTIVE

To identify operations for which mechanical exhaust ventilation may be needed, and provide information on design parameters and criteria for evaluating system performance.

BACKGROUND

Ventilation is a widely applied method for maintaining an acceptable temperature of a work area (comfort ventilation) or for maintaining airborne contamination at acceptable levels. This is achieved by providing general ventilation to the area, preventing the release of contaminants from the task being carried out by use of local exhaust ventilation, or a combination of both. Air must be provided to make up that removed from a space by a local exhaust or general ventilation system.

General ventilation is the practice of supplying air to an area by mechanical means, removing air by mechanical means, and providing make-up air to replace that exhausted, or by providing for natural ventilation. The latter is dependent on wind pressure and thermal differences (gravitational forces) and, as would be expected, is unreliable for contamination control.

General ventilation is the method of choice to provide for the comfort of personnel by heating or cooling the air that is supplied to an area. This method has limited application for airborne contamination control and may only be considered for this application when the toxicity of the contaminant of concern is low, the rate of generation is small and reasonably constant, personnel work at a distance from the point of release, and an appropriate dilution-effectiveness (mixing) factor is applied to the supply air rate. Other considerations in the use of general ventilation for contamination control include the potential for seasonal effects, number of points of contaminant generation, ventilation system maintenance effectiveness, etc. Areas where general ventilation has been used for contamination control include pump rooms on ships, welding areas, lube blending and grease making facilities, chemical storage areas, sample retain rooms, and others (see Table 1).

Local exhaust ventilation systems are those which minimize the release of contaminant from a source by containing or capturing it at or near the point of generation and removing it from the work area. These systems include an enclosure, such as a hood, ductwork through which air is transported from the enclosure through an air cleaning device (if warranted) to an air mover, then through exhaust ductwork for discharge to the outside air.

Local exhaust ventilation systems are the method generally selected when the objective is airborne contamination control for health protection. Advantages of local exhaust systems include their effectiveness for capturing and removing airborne contaminants with consequent reduction in exposure to the contaminant; reduced exhaust volumes resulting in significant energy savings; smaller investment for system components; and reduced cost for exhaust air clean-up, when necessary. Operations at which local exhaust systems are employed include the laboratory, spray painting, grinding, welding/metallizing, drum filling, mixing/blending, bag filling, plastic-film treating, and others.

Local exhaust systems must have the appropriate control velocity (prevent release of contaminant from the hood) or capture velocity (air movement at point of contaminant generation that is sufficient to capture and convey the contaminant into the hood) to be effective. The air velocity necessary for exposure control depends upon the characteristics of contaminant generation (temperature, kinetic energy, etc.), the direction and magnitude of air motion surrounding the point of contaminant generation, the position of the work with respect to hood location and the point of contaminant generation, and whether adequate make-up air is supplied to the general area in which the local exhaust system is operated. All of these factors must be considered in designing or evaluating the performance of a local exhaust ventilation system.

CONSIDERATIONS

Information must be obtained regarding the operation for which a ventilation system is to be provided so that a proper design can be developed. The properties of the material to be controlled must be known so that an effective hood, exhaust rate, air cleaner, and exhaust discharge location can be selected and specified. Special ductwork, motor, and fan material may be required. The location in which the system will be installed and operated should be evaluated to identify potential limitations, if any, which could affect installation.

The design should be reviewed by the appropriate functional group or groups, possibly including engineering, industrial hygiene, safety, environmental, etc. to ensure it will be adequate. Some of the parameters that should be evaluated include the type of hood selected; air velocity needed for contamination control or capture; ductwork (size, elbows, entries, inspection/clean-out ports, etc); fan type and construction; type of stackhead; point of discharge relative to occupied areas and the location of air inlets with the potential for re-entry of contaminants; adequacy of make-up air; and others. After the ventilation system has been installed, the system should be evaluated to ensure it meets design criteria, achieves effective control, and contaminant re-entry does not occur. The data generated in this effort should be retained and filed for future reference (e.g., as baseline data) in the periodic re-evaluation of the system.

REFERENCES

"American National Standard Fundamentals Governing the Design and Operation of Local Exhaust Systems" ANSI-Z9.2 (New York: American National Standards Institute, 1979).

"American National Standard Methods of Laboratory Air-Flow Measurement" ANSI-ASHRAE (New York: American National Standards Institute, 1979).

"American National Standard for Exhaust Systems - Spray Finishing Operations", ANSI/ASC-Z9.3 (New York: American National Standards Institute, 1985).

"American National Standard for Exhaust Systems - Abrasive Blasting Operations - Ventilation and Safe Practices", ANSI-Z9.4 (New York: American National Standards Institute, 1985).

Burgess, W. A., M. J. Ellenbecker and R. T. Trietman, "Ventilation for Control of the Work Environment" (New York: John Wiley & Sons, Inc., 1989).

Di Berardinis, L. J., M. W. First and R. E. Ivany, "Field Results of an In-Place, Quantitative Performance Test for Laboratory Fume Hoods", Applied Occupational and Environmental Hygiene, March 1991, pp. 227-231.

"Industrial Ventilation - A Manual of Recommended Practice", Latest Edition (Lansing, MI: American Conference of Governmental Industrial Hygienists).

Soule, R. D., "Industrial Hygiene Engineering Controls", Vol. 1, General Principles, Patty's Industrial Hygiene and Toxicology (New York: John Wiley & Sons, Inc.), pp. 771-823.

TABLE I

TYPICAL PETROLEUM/PETROCHEMICAL OPERATIONS WHICH HAVE VENTILATION CONTROLS

OPERATION	TYPE OF VENTILATION SYSTEM EMPLOYED	
	GENERAL	LOCAL EXHAUST
Abrasive Cleaning	X (Area)	X (Hood)
Bag Dumping (Filter Aid)		X (Hood)
Blending Tanks	X (Area)	X (Tank)
Carpenter Shop/Maintenance Shop	X (Area)	X (Hood)
Control Room (Supply Air to Room)	X (Area)	
Degreasing	X (Area)	X (Hood)
Drum Cleaning/Reconditioning	X (Area)	X (Hood)
Drum Filling	X (Area)	X (Drum)
Grease Making	X (Area)	X (Kettles)
Instrument Shed	X (Shed)	
Laboratory	X (Area)	X (Hoods)
LPG Cylinder Filling Facility:		
Abrasive Cleaning		X (Cabinet)
Spray Painting		X (Hood)
Filling	X (Area)	
Lube Blending	X (Area)	X (Tanks)
Metallizing		X (Hood)
Plate and Frame Filter	X (Area)	X (Hood)
Pump Room	X (Area)	
Reactors:		
Catalyst Addition	X (Area)	
Catalyst Dumping	X (Area)	X (Hood)
Rotary Filter		X (Enclosed)
Sampling Process Streams	X (Area)	X (Hood)
Spray Painting	X (Area)	X (Hood)
Storage Room	X (Area)	
Sub-Station (Supply Air to Enclosure)	X (Area)	
Tank Cleaning	X (Top Opening and Manway)	
Welding	X (Area)	X (Hoods)

WORKPLACE HEALTH PROTECTION PROGRAM ELEMENT CHECKLIST ON VENTILATION

	Yes	No	Don't Know	N/A
1. Are pre-construction ventilation system design reviews routinely performed by:				
. Engineering?	___	___	___	___
. Environmental?	___	___	___	___
. Industrial hygiene?	___	___	___	___
. Safety?	___	___	___	___
. Other? (________________________)	___	___	___	___
2. Is ventilation system performance evaluated after installation is completed?	___	___	___	___
3. Is ventilation system performance evaluated periodically (e.g., annually for laboratory fume hoods)?	___	___	___	___
4. Are records of ventilation system performance documented and retained for comparison to results of future evaluations?	___	___	___	___
5. Are the following satisfactory for each local exhaust system:				
. Hood type?	___	___	___	___
. Face (control) velocity?	___	___	___	___
. Capture velocity?	___	___	___	___
. Transport velocity?	___	___	___	___
. Make-up air?	___	___	___	___
. Ductwork construction:				
- Proper for contaminant?	___	___	___	___
- Proper elbow design?	___	___	___	___
- Good tapered entries?	___	___	___	___
- Clean-out ports?	___	___	___	___
. Fan construction?	___	___	___	___
. Fan location?	___	___	___	___
. Air cleaner?	___	___	___	___
. Stackhead design?	___	___	___	___

N/A - Not Applicable

WORKPLACE HEALTH PROTECTION PROGRAM ELEMENT CHECKLIST ON VENTILATION

	Yes	No	Don't Know	N/A
6. Are local exhaust systems adequate at:				
. Laboratory hoods?	___	___	___	___
. Glassware wash sink?	___	___	___	___
. Welding hoods?	___	___	___	___
. Blending tanks?	___	___	___	___
. Grease kettles?	___	___	___	___
. Paint spray booths?	___	___	___	___
. Film treaters?	___	___	___	___
. Other? (________________________)	___	___	___	___
7. Is general ventilation adequate for airborne contamination control at the:				
. Lead blending facility?	___	___	___	___
. Grease making facility?	___	___	___	___
. Laboratory:				
- General area?	___	___	___	___
- Knock test laboratory?	___	___	___	___
- Sample retain room?	___	___	___	___
- Glassware wash room?	___	___	___	___
. Lube blending area?	___	___	___	___
8. Are laboratory fume hoods provided with flow indicators that are visible to the user?	___	___	___	___
9. Are indicators provided on ventilation equipment to demonstrate operating effectiveness:				
. Paint spray booths?	___	___	___	___
. Bag collectors?	___	___	___	___
. Ventilation system filters?	___	___	___	___
10. Is the general air supply for laboratories (in multi-use buildings) designed to flow from offices to corridors to laboratory space?	___	___	___	___
11. Has the potential for hood exhaust re-entry been minimized?	___	___	___	___
12. Are canopy type hoods (e.g., above atomic absorption spectrometers, filter presses, etc.) effective for capturing and removing contaminants?	___	___	___	___

N/A - Not Applicable

WORKPLACE HEALTH PROTECTION PROGRAM ELEMENT CHECKLIST
ON
VENTILATION

	Question	Yes	No	Don't Know	N/A
13.	Are exhaust fans equipped with alarms to indicate if they fail?	___	___	___	___
14.	Are laboratory hoods provided with an airfoil design?	___	___	___	___
15.	Is supply air velocity in work areas sufficiently low to prevent drafts on personnel?	___	___	___	___
16.	Are exhaust hoods located away from supply air grills so that turbulence at the hood face is minimized?	___	___	___	___
17.	Have personnel been trained on how to use and work at laboratory hoods?	___	___	___	___
18.	Is available ventilation measuring equipment:				
	. The appropriate type for use in the facility?	___	___	___	___
	. Periodically calibrated to ensure accuracy?	___	___	___	___
19.	Is general ventilation adequate to prevent excessive heat stress at:				
	. Furnace inspection operations?	___	___	___	___
	. Lube blending operations?	___	___	___	___
	. Grease making operations?	___	___	___	___
	. Tank cleaning operations?	___	___	___	___
	. Other? (________________)	___	___	___	___
20.	Are welding fume hoods effectively used by personnel?	___	___	___	___

N/A - Not Applicable

4. SUPPORT ACTIVITIES

WORKPLACE HEALTH PROTECTION PROGRAM ELEMENT
ON
ACQUISITIONS

WORKPLACE HEALTH PROTECTION PROGRAM ELEMENT ON ACQUISITIONS

OBJECTIVE

To provide a means to identify the workplace health issues that management should be aware of when considering the acquisition of a facility or company.

BACKGROUND

The acquisition of a facility may result in the company assuming liability for adverse health effects among personnel which preceded ownership, as well as for effects which may arise in the future as a result of inadequate procedures/work practices adhered to in the past. For example, if personnel had been excessively exposed to a chronic toxicant, such as asbestos, the potential adverse health effect may not, as yet, be evident or diagnosed.

Other issues of concern may encompass the broad spectrum of adverse occupational outcomes. Hearing loss may exist among employees and not be known if the previous employer did not have an effective audiometric testing program. Use of silica containing materials in abrasive cleaning operations may have resulted in the excessive exposure of personnel to silica containing particulates with the consequent potential for their developing silicosis. Often, abrasive cleaning is done on surfaces that have been painted with lead or chromate based paints. These contaminants can give rise to adverse health effects that may not have been detected if the previous medical surveillance program was not directed toward detection of such effects.

A listing of materials and amounts used in operations, as well as an exposure assessment review, should be conducted to identify workplace health concerns. Information developed in this effort can have a bearing on the decision as to whether to proceed with the acquisition. It can also provide information on potential liability, as well as identify program needs that should be implemented if the purchase is made. However, a pre-purchase inventory will identify only those materials currently used and not those used in the past but which are no longer in use.

CONSIDERATIONS

Management should assemble a team of health professionals and other personnel to assess facility conditions, exposure risks, procedures, work practices, and possible industrial hygiene program needs that may have to be implemented if a decision is made to acquire the facility or company. Team members should include a physician, toxicologist, safety professional, environmental health specialist, and an industrial hygienist to assess relevant issues, including workplace health concerns that are associated with operations.

Management should arrange a visit and walk-through inspection by the team so a report can be developed identifying the status of the occupational health, safety, environmental, and industrial hygiene programs that are in place or needed. It is essential that process information and a detailed inventory of materials used in the operation, as well as a description of the process be available to the team before the walk-through inspection is undertaken.

The present owner should be requested to provide information on citations by regulatory agencies relative to workplace health related issues. Medical records of injuries and occupational diseases should be reviewed. Survey reports, monitoring data, and findings of previous program audits should be reviewed, as well as the findings of follow-up studies assessing the effectiveness of corrective actions taken to comply with recommendations. In addition, management should be made aware of outstanding items. Information should be obtained relative to community reaction/complaints that have arisen due to operations. It would be advisable to obtain all relevant facts for management consideration if there have been such occurrences.

Existing health protection programs should be evaluated relative to company guidelines and applicable regulations. The effort to achieve compliance should be determined. When all pertinent information has been obtained, a report (written or verbal per management's direction) should be prepared and submitted to management.

WORKPLACE HEALTH PROTECTION PROGRAM ELEMENT CHECKLIST ON ACQUISITIONS

	Yes	No	Don't Know	N/A
1. When considering a new acquisition, is a review procedure implemented which addresses workplace health protection issues?	___	___	___	___
2. Has the potential seller been requested to provide:				
. Process information?	___	___	___	___
. Inventory of materials present/used in the facility?	___	___	___	___
. MSDSs of:				
- Materials used?	___	___	___	___
- Intermediates?	___	___	___	___
- Products?	___	___	___	___
- Wastes generated?	___	___	___	___
3. Are copies of previous industrial hygiene surveys available?	___	___	___	___
4. Have workplace health protection survey reports been reviewed?	___	___	___	___
5. Has monitoring been conducted to determine worker's exposure to:				
. Chemical agents?	___	___	___	___
. Physical agents?	___	___	___	___
. Biologic agents?	___	___	___	___
. Ergonomic factors?	___	___	___	___
6. Is worker exposure data available?	___	___	___	___
7. Have exposure levels of hazardous substances and physical agents been acceptable?	___	___	___	___
8. Have monitoring results been made available to employees?	___	___	___	___
9. Have exposures to physical, chemical, and biological agents been effectively controlled, where necessary, to reduce the potential for adverse health effects to occur?	___	___	___	___

N/A - Not Applicable

WORKPLACE HEALTH PROTECTION PROGRAM ELEMENT CHECKLIST ON ACQUISITIONS

		Yes	No	Don't Know	N/A
10.	Has an asbestos-containing material (ACM) inventory been carried out in the facility?	___	___	___	___
11.	Is asbestos-containing material (ACM) present:				
	. Equipment/piping insulation?	___	___	___	___
	. Sprayed-on insulation?	___	___	___	___
	. Gaskets?	___	___	___	___
	. Pump/compressor seals?	___	___	___	___
	. Ceiling tile?	___	___	___	___
	. Floor tile?	___	___	___	___
	. Fireproofing material?	___	___	___	___
	. Other building materials (e.g., asbestos concrete)?	___	___	___	___
12.	Is friable ACM present?	___	___	___	___
13.	Are asbestos/MMMF-containing materials in an acceptable condition?	___	___	___	___
14.	If ACM is present is there a written operations and maintenance program in place?	___	___	___	___
15.	Has any asbestos abatement been carried out in the facility?	___	___	___	___
16.	Based on observations made during a walk-through of the facility, is exposure potential considered acceptable for:				
	. Fibers (asbestos, MMMF)?	___	___	___	___
	. Physical agents (noise, heat, etc.)?	___	___	___	___
	. Vapors (hydrocarbons, benzene, etc.)?	___	___	___	___
	. Fumes (metal, asphalt, etc.)?	___	___	___	___
	. Dust (coke, silica, etc.)?	___	___	___	___
	. Gases (chlorine, ammonia, etc.)?	___	___	___	___
	. Ergonomic aspects?	___	___	___	___
	. Other (________________)?	___	___	___	___

N/A - Not Applicable

WORKPLACE HEALTH PROTECTION PROGRAM ELEMENT CHECKLIST ON ACQUISITIONS

		Yes	No	Don't Know	N/A
17.	Is the use of respiratory protection unnecessary at routine operations?	___	___	___	___
18.	Has the facility experienced any emergency situations which resulted in:				
	. Unacceptable releases to the environment?	___	___	___	___
	. Excessive exposure of emergency forces?	___	___	___	___
	. Contamination of the area?	___	___	___	___
	. Community complaints?	___	___	___	___
	. Excessive employee exposures?	___	___	___	___
	. Evacuation of personnel?	___	___	___	___
19.	Has the potential for the occurrence of workplace health related emergencies been addressed and reduced?	___	___	___	___
20.	Has a comprehensive workplace health protection program review ever been conducted?	___	___	___	___
21.	Has PCB-containing equipment been identified?	___	___	___	___
22.	Have leaks from PCB containing equipment been prevented?	___	___	___	___
23.	Have measures been taken to remove PCB's from equipment?	___	___	___	___
24.	Is there a safety handbook on site?	___	___	___	___
25.	Is there a safety committee?	___	___	___	___
26.	Does the safety committee address workplace health protection issues?	___	___	___	___

N/A - Not Applicable

WORKPLACE HEALTH PROTECTION PROGRAM ELEMENT CHECKLIST ON ACQUISITIONS

		Yes	No	Don't Know	N/A
27.	Are the work procedures/work practices that are defined in the safety handbook (i.e., those related to workplace health issues) adequate for the associated potential exposures?	___	___	___	___
28.	Are employees aware of the health hazards associated with the substances used and produced?	___	___	___	___
29.	Is signage (i.e., health hazard signs) adequate throughout the facility?	___	___	___	___
30.	Is appropriate personal protective equipment available and used as needed?	___	___	___	___
31.	Is emergency equipment (i.e., for health protection purposes) in good condition throughout the facility?	___	___	___	___
32.	Have personnel been trained to respond to emergency situations?	___	___	___	___
33.	Are the following program elements adequate:				
	. Asbestos?	___	___	___	___
	. Confined space entry?	___	___	___	___
	. Continuous monitors?	___	___	___	___
	. Ergonomics?	___	___	___	___
	. Hazard awareness?	___	___	___	___
	. Hearing conservation?	___	___	___	___
	. Hearing protection?	___	___	___	___
	. Lighting?	___	___	___	___
	. Monitoring?	___	___	___	___
	. Permits?	___	___	___	___
	. Personal protection?	___	___	___	___
	. Radiation protection?	___	___	___	___
	. Recordkeeping?	___	___	___	___
	. Respiratory protection?	___	___	___	___
	. Ventilation?	___	___	___	___
	. Others (________________)?	___	___	___	___

N/A - Not Applicable

WORKPLACE HEALTH PROTECTION PROGRAM ELEMENT CHECKLIST ON ACQUISITIONS

		Yes	No	Don't Know	N/A
34.	Have biological tests been carried out on personnel, as appropriate (e.g., for hearing, lung function, lead in blood/urine, etc.)?	___	___	___	___
35.	Have results of biological testing of personnel been acceptable (i.e., those related to workplace health issues)?	___	___	___	___
36.	Has occupational hearing loss been prevented among employees?	___	___	___	___
37.	Are ionizing radiation sources properly licensed, registered, identified, and used?	___	___	___	___
38.	Are ionizing radiation source records complete?	___	___	___	___
39.	Have any ionizing radiation sources been disposed on site?	___	___	___	___
40.	Have any process or other wastes been disposed on site?	___	___	___	___
41.	Have potential incompatibilities been addressed in chemical storage areas?	___	___	___	___
42.	Are laboratory ventilation systems effective:				
	. Hoods (shop, laboratory, etc.)?	___	___	___	___
	. General ventilation?	___	___	___	___
	. Bottle-wash facility?	___	___	___	___
	. Storage areas?	___	___	___	___
	. Other (_________________)?	___	___	___	___
43.	Are products, intermediates, feedstocks, lab chemicals, utility chemicals, samples, etc. properly labeled?	___	___	___	___

N/A - Not Applicable

WORKPLACE HEALTH PROTECTION PROGRAM ELEMENT CHECKLIST ON ACQUISITIONS

		Yes	No	Don't Know	N/A
44.	Have processes/operations that were previously carried out on the site/in the facility (but no longer are) been identified?	___	___	___	___
45.	Are there monitoring results for radon levels in the facility?	___	___	___	___
46.	Has abatement been unnecessary to reduce radon contamination levels?	___	___	___	___
47.	Have complaints of poor indoor air quality been addressed?	___	___	___	___
48.	Have indoor air quality problems been corrected?	___	___	___	___

N/A - Not Applicable

WORKPLACE HEALTH PROTECTION PROGRAM ELEMENT
ON
CONTRACTOR SUPPORT

WORKPLACE HEALTH PROTECTION PROGRAM ELEMENT ON CONTRACTOR SUPPORT

OBJECTIVE

To provide information on the workplace health protection issues associated with work being carried out by contractors and identify means to control exposure to the potential health hazards associated with their activities.

BACKGROUND

Contractors are independent business people or organizations retained to perform work for an agreed upon price. Often, sub-contractors are hired to perform a portion of the work specified in a contract that has been awarded a prime contractor.

It is common practice to hire contractors to do some of the work in an industrial facility. Maintenance and construction are two of the work activities that contractors are typically called upon to carry out. It is essential that contractors and their employees to aware of the potential health hazards associated with work in a facility, as well as the required procedures and work practices to be followed.

It is necessary that effective lines of communication be established between facility and contractor management before the contract is let, and that the communication be continued throughout the period of the work. Information should be provided to, and obtained from, contractors before a successful bidder is selected to do the contract work. The workplace health hazard information provided to contractor management must be transmitted to the contractor's work force. It is also the responsibility of the contractor management to ensure that their employees comply with established facility procedures/work practices and carry out their work without endangering their health or that of others.

CONSIDERATIONS

Facility management should develop procedures to be followed as a part of the contractor selection process. This could include a meeting with prospective contractors to outline the details of the work to be done and the associated health/environmental concerns, a review of bid specifications and bid submittals by responsible health/safety personnel, as well as the transmittal of relevant workplace health hazard information to potential bidders. A request for similar information from the contractor should be made for inclusion as part of the bid submission.

Information to be provided to the contractor by facility management with regard to workplace health protection issues could include:

- When appropriate, a listing of governmental agencies that have regulations which will impact the contractor's work (federal, province, state, local, etc.)
- A listing of the unusual hazardous materials and potential health hazards the contractor or sub-contractor may encounter in doing the work (e.g., lead, asbestos, benzene, noise, etc.) and of the alarm systems in place to alert personnel of a hazardous situation and the action to take if an emergency arises.
- A clear statement indicating that the contractor is responsible for ensuring a safe and healthful work environment for contractor and sub-contractor personnel.
- Relevant information (e.g., MSDSs) on materials to which contractor/sub-contractor personnel may be exposed.
- Information on rules, standards, work practices, permit requirements, special procedures, etc. that the contractor/sub-contractor will need to comply with in carrying out work in the facility (e.g., such as that presented in the site safety handbook).
- The need for special protective equipment that will be required in carrying out the work.
- The name of a contact who is available to discuss issues related to workplace health protection program considerations that may be applicable for the work.
- Restrictions on the use of facility services (locker rooms, showers, cafeteria, etc.).
- Management guides on specific workplace health related issues that may be applicable to the work to be carried out.

The contractors should be informed of the requirement that:

- They are to restrict their personnel to the designated work site and to maintain satisfactory housekeeping in the work area.
- All waste materials generated in the work area are to be disposed in accordance with applicable regulations.
- All required protective equipment for their employees will be provided by the contractor.
- They provide the facility with copies of accident investigations, spill/air release reports, etc.

- They are to conduct appropriate monitoring of their personnel to comply with regulations and company requirements.

Information to be obtained from the contractor as a part of the bid submission may include:

- Details of the contractor's workplace health policy and protection program.

- Copies of the contractor's most recent government inspection reports relative to their workplace health protection program (when available and applicable).

- Copies of employee training records (as related to requirements associated with contracted work).

- Material safety data sheets on hazardous materials to be brought on site and used by the contractor/sub-contractors. These should be reviewed by a responsible facility individual before they are allowed to be brought on site.

- Documentation that the contractor and sub-contractor will be responsible for:

  - All necessary training of contractor and sub-contractor personnel.
  - Providing all personal protective equipment to contractor and sub-contractor employees.
  - Effective disposal of wastes generated in carrying out the work.
  - Providing reports of injuries, spills, waste disposal, etc. to facility management.
  - Periodically (e.g., daily) inspecting the work area to ensure that health hazards are effectively controlled. Facility personnel could also conduct similar inspections.
  - Reviewing the workplace health protection procedures/practices of their personnel to ensure compliance with applicable regulations and Mobil guidelines.
  - Providing facility management a copy of the monitoring results.

- Agreement that the contractor and sub-contractors will enable facility personnel to review their work activities with respect to workplace health protection program considerations, carry out inspections of the work area and, when appropriate, interact with contract supervision to take action to control a hazard that may exist or develop in the course of the work.

- Facility management should maintain a listing of acceptable contractors.

WORKPLACE HEALTH PROTECTION PROGRAM ELEMENT CHECKLIST ON CONTRACTOR SUPPORT

		Yes	No	Don't Know	N/A
1.	Is information provided to prospective contract bidders regarding workplace health protection issues related to the work to be done as a part of the bidding procedure?	___	___	___	___
2.	Does the contractor have a written workplace health protection policy and program?	___	___	___	___
3.	Does the contractor have a written workplace health/safety protection manual?	___	___	___	___
4.	Is a facility contact identified to prospective contract bidders as an information source regarding workplace health protection matters?	___	___	___	___
5.	Is a copy of the site safety manual provided to, or made available to prospective contractors?	___	___	___	___
6.	When appropriate, are copies of applicable management guides on workplace health related issues made available to potential contractors?	___	___	___	___
7.	Is information transmitted to potential contractors regarding:				
	. A general description of potential health hazards that may be associated with the contracted work?	___	___	___	___
	. Contractor responsibility for ensuring sub-contractors comply with regulations, as well as required procedures/work practices?	___	___	___	___
	. Special procedures and work practices that are required?	___	___	___	___
	. Contaminants to which contractor/sub-contractor employees may be exposed during their work in the facility?	___	___	___	___

N/A - Not Applicable

WORKPLACE HEALTH PROTECTION PROGRAM ELEMENT CHECKLIST
ON
CONTRACTORS SUPPORT

	Yes	No	Don't Know	N/A
. List of restrictions for contractor use of facilities?	___	___	___	___
8. Has the contractor been advised of the need to provide:				
. Hazard communication training for their personnel?	___	___	___	___
. Personal protective equipment training?	___	___	___	___
. Non-asbestos containing insulation/ gaskets/pump seals/etc.?	___	___	___	___
. Effective disposal of wastes generated in doing the contracted work?	___	___	___	___
. Reports of injuries and spills/releases to facility management and regulatory agencies?	___	___	___	___
. Monitoring of personnel to meet regulatory requirements?	___	___	___	___
. Copies of the monitoring results to facility management?	___	___	___	___
. Material safety data sheets on materials brought into the facility?	___	___	___	___
. Facility personnel an opportunity to review contractor/sub-contractor work activities with respect to workplace health protection issues?	___	___	___	___
. Good housekeeping in the work area?	___	___	___	___
9. Are bid submissions reviewed by responsible groups with respect to workplace health protection issues?	___	___	___	___
10. Does the contract clearly indicate that the contractor is responsible for providing contract workers with safe and healthful working conditions?	___	___	___	___
11. Does the contractor have a procedure in place to ensure that sub-contractor personnel are aware of the workplace health hazards and appropriate exposure control procedures?	___	___	___	___

N/A - Not Applicable

WORKPLACE HEALTH PROTECTION PROGRAM ELEMENT CHECKLIST ON CONTRACTOR SUPPORT

	Yes	No	Don't Know	N/A
12. Is information obtained from the contractor/sub-contractor regarding:				
. Their workplace health protection program?	___	___	___	___
. Results (where applicable) of the most recent governmental inspection regarding their workplace health protection program?	___	___	___	___
. Employee training records (related to workplace health protection issues)?	___	___	___	___
. Material or equipment that will be brought onto the site that may pose a health hazard to personnel?	___	___	___	___
13. Have reference documents on other workplace health protection program elements been reviewed to ensure the procedures and work practices of a contractor are acceptable with respect to:				
. Abrasive cleaning work?	___	___	___	___
. Asbestos work?	___	___	___	___
. Confined space entry?	___	___	___	___
. Emergency response?	___	___	___	___
. Hazard awareness training?	___	___	___	___
. Hearing protection use?	___	___	___	___
. Other protective equipment use?	___	___	___	___
. Permit compliance?	___	___	___	___
. Radiography (Ionizing Radiation)?	___	___	___	___
. Respirator use (Respiratory Protection)?	___	___	___	___
. Tank Cleaning?	___	___	___	___
. Ventilation?	___	___	___	___
. Other? (________________)	___	___	___	___
14. Are MSDSs of materials to be used by contractor/sub-contractor reviewed by a responsible facility group before they are brought onto the site?	___	___	___	___

N/A - Not Applicable

WORKPLACE HEALTH PROTECTION PROGRAM ELEMENT CHECKLIST ON CONTRACTOR SUPPORT

		Yes	No	Don't Know	N/A
15.	Are written compliance programs of the contractor/sub-contractor reviewed by responsible site personnel?	___	___	___	___
16.	Are contractor workplace health protection training programs evaluated?	___	___	___	___
17.	Is contractor training documentation provided/reviewed/satisfactory?	___	___	___	___
18.	Is the contractor's workplace health protection policy communicated to employees/sub-contractor?	___	___	___	___
19.	Does the contractor have industrial hygiene support?	___	___	___	___
20.	Do responsible facility personnel provide advice to contractors on MSDSs, as requested?	___	___	___	___
21.	Has emergency response information been transmitted to contractor/sub-contractor employees?	___	___	___	___
22.	Has the contractor been made aware of the facility's emergency:				
	. Alarms signals?	___	___	___	___
	. Procedures?	___	___	___	___
	. Assembly area for personnel?	___	___	___	___
23.	Do contractors/sub-contractors keep their personnel aware of potential exposure risks associated with work being carried out?	___	___	___	___
24.	Does the industrial hygiene function interact with the contractor regarding workplace health protection issues?	___	___	___	___
25.	Do responsible site personnel periodically evaluate contractor/sub-contractor work practices?	___	___	___	___

N/A - Not Applicable

WORKPLACE HEALTH PROTECTION PROGRAM ELEMENT CHECKLIST ON CONTRACTOR SUPPORT

	Question	Yes	No	Don't Know	N/A
26.	Do responsible site personnel review contractor/sub-contractor monitoring data?	___	___	___	___
27.	Do responsible site personnel review permits to ensure compliance with requirements related to workplace health protection issues?	___	___	___	___
28.	Is there assurance that incidents (contractor/sub-contractor) related to workplace health protection program issues are reported to facility management?	___	___	___	___
29.	Are regular workplace health protection inspections of the workplace carried out by a trained contractor employee/consultant?	___	___	___	___
30.	Does a facility representative (e.g., environmental, industrial hygiene, safety, etc.) have authority to stop a job for cause (i.e., related to workplace health issues)?	___	___	___	___
31.	Can the contract be terminated in the event of repeated or serious non-compliance with accepted/established workplace health protection program procedures/practices?	___	___	___	___
32.	Can the contractor(s) be removed from an approved list if their workplace health protection program is determined to be inadequate?	___	___	___	___
33.	Does the contractor provide adequate personal hygiene facilities for contractor/sub-contractor personnel?	___	___	___	___

N/A - Not Applicable

WORKPLACE HEALTH PROTECTION PROGRAM ELEMENT
ON
FOLLOW-UP OF RECOMMENDATIONS

WORKPLACE HEALTH PROTECTION PROGRAM ELEMENT
ON
FOLLOW-UP OF RECOMMENDATIONS

OBJECTIVE

To identify the need for prompt follow-up after recommendations have been made and implemented and ensure that the action taken has effectively reduced the health hazard.

BACKGROUND

If employee exposure to chemical, physical, biological, or ergonomic stress is excessive, recommendations are made to reduce the hazard. Measures to be taken for reducing exposure risk will depend upon the properties of the offending agent and the magnitude of the associated hazard. Corrective action may include:

- Reviewing and revising the procedures/work practices as necessary
- Educating potentially exposed personnel
- Providing sufficient general ventilation to dilute the offending agent to an acceptable level
- Providing adequate local exhaust ventilation to remove the offending agent from the workplace
- Repositioning equipment
- Isolating the source of the offending agent from the work area
- Isolating excessively exposed individuals from the contaminated environment
- Replacement of equipment
- Substituting the offending agent with a less hazardous one
- Relocating an operation to a non-occupied area
- Providing personal protective equipment to personnel
- Mechanizing the operation
- Discontinuing the operation

The decision as to when corrective action will be taken and which method to employ for reducing the exposure risk rests with management. It is the responsibility of the health professional, as well as others, to provide advice on which control method would be most effective for the specific situation. There must be follow-up after the exposure control method has been implemented to ensure that the hazard has been sufficiently reduced.

CONSIDERATIONS

A report of findings of each workplace health risk assessment (survey, investigation, etc.), should be forwarded to management. Management should identify the functional group that will be responsible for the corrective action to be taken to reduce the hazard. This group should identify the estimated date of completion of the control measure. This information will enable scheduling a follow-up assessment to determine the exposure control methodology and ensure that the hazard has been effectively reduced. When controls are effective for reducing the hazard, a report of findings should be sent to management and a copy filed. If controls are found not to be effective, a meeting should be requested with management to discuss the findings and identify additional measures to be taken. Follow-up evaluations should be continued until exposure to the hazard is controlled.

WORKPLACE HEALTH PROTECTION PROGRAM ELEMENT CHECKLIST ON FOLLOW-UP OF RECOMMENDATIONS

		Yes	No	Don't Know	N/A
1.	Is management provided a report of findings of a survey/ investigation which was the basis for recommendations to reduce a workplace health hazard?	___	___	___	___
2.	Is management requested to identify the group responsible for:				
	. Implementing control measures to reduce a workplace health hazard?	___	___	___	___
	. Periodic reports of control implementation status?	___	___	___	___
	. Providing a projected date for completion of control implementation?	___	___	___	___
3.	Are follow-up evaluations scheduled and carried out to determine the effectiveness of controls that have been implemented in response to recommendations made?	___	___	___	___
4.	Does documentation exist which indicates that recommendations made for reducing workplace health hazards have been implemented and the hazards have been satisfactorily controlled?	___	___	___	___
5.	Have affected employees been advised of the findings that were the basis for an exposure control recommendation?	___	___	___	___
6.	Is there an ongoing effort to achieve effective health hazard control if the measures taken have not reduced the hazard satisfactorily?	___	___	___	___

N/A - Not Applicable

WORKPLACE HEALTH PROTECTION PROGRAM ELEMENT CHECKLIST ON FOLLOW-UP OF RECOMMENDATIONS

		Yes	No	Don't Know	N/A
7.	Is use of protective equipment required until an identified health hazard has been effectively reduced by engineering/administrative means?	___	___	___	___
8.	Are affected employees informed when the measures taken have effectively controlled an exposure hazard?	___	___	___	___

N/A - Not Applicable

WORKPLACE HEALTH PROTECTION PROGRAM ELEMENT ON HAZARD AWARENESS

WORKPLACE HEALTH PROTECTION PROGRAM ELEMENT ON HAZARD AWARENESS

OBJECTIVE

To identify the need to provide periodic awareness training for facility personnel and others potentially exposed to workplace health hazards.

BACKGROUND

Workplace health hazards include any material, physical agent, biological organism or ergonomic stress for which there is evidence that acute or chronic health effects may result from exposure to it. Included are carcinogens, reproductive toxins, irritants, corrosives, sensitizers, noise, vibration, heat stress, ionizing and non-ionizing radiation, repeated stress, pathogenic organisms and others. Employees may be exposed to some or all of these in the course of their work.

The potential adverse health effects that may result as a consequence of excessive exposure to workplace health hazards are generally well known among occupational health personnel. In addition, methods for reducing exposure risk have been developed, whether through the application of engineering controls, administrative measures, use of personal protective equipment or a combination of these. When not known, appropriate exposure control information may be obtained from a review of the literature, from trade associations, through contacts with other occupational health specialists, or from other sources.

Management has overall responsibility to ensure that employees are not excessively exposed to the health hazards that may be present in the workplace. To accomplish this, supervisory personnel have responsibility to implement procedures/work practices and effective exposure control measures and ensure that personnel use them.

In some instances, employees have been reluctant to follow established work practices or to use available control measures. This may have been due to inadequate training as to the reasons for use of the exposure control measures and/or not being aware of the adverse health effect potential resulting from exposure to a hazardous agent. It is not sufficient that only health professionals and management be aware of the health hazards associated with operations. Employees too, need to be aware of the potential hazards to which they may be exposed, what the effects of these are, and what they can do to prevent or reduce their exposure to the hazardous agent(s). Such information should be provided to employees in periodic hazard awareness training programs.

CONSIDERATIONS

A written hazard awareness training program should be developed and implemented. It should be presented to management personnel and all employees who may be exposed to hazardous materials, biological agents, physical agents or ergonomic stresses in the course of their work. The training should be repeated periodically. In addition, when a new material is to be used, a process change is to be made, or an individual is to be transferred to a new area/job that may present a new hazard, appropriate health hazard awareness training should be provided.

Information presented in the hazard awareness program should include training to enable personnel to identify hazardous materials in the work area; how to work with these without excessive risk (i.e., use of proper procedures/work practices, protective equipment, etc.); how to read and understand a material safety data sheet; an understanding of labeling practices; the adverse health effects that may develop as a result of excessive exposure to a hazard; where additional information may be obtained; where exposure to the hazard may occur; and information on how exposure risk is evaluated.

Responsibility for hazard awareness program development should be assigned to an individual or function. The same applies to hazard awareness training, although not necessarily to the same individual or function that has responsibility for program development. These activities should be stewarded to management and records maintained of those who have received the training.

The information to be presented in the training should be related to the operations being carried out, and to the inventory of materials received, handled, used, or produced in the facility. Material safety data sheets should be available to personnel and they should receive instructions on their interpretation. Information on physical agents such as noise, radiation and other physical health hazards needs to be included in the training.

An individual or function should be responsible for reviewing new material requisitions to ensure appropriate controls will be in place for their use. In addition, the material safety data sheets for each new material requisitioned should be reviewed for completeness when received. Periodic reviews can be carried out to determine if MSDSs are available in all areas of material use. In addition, workers must be aware of the potential hazards, and know how to use the hazard information that is to be made available to them. Personnel working in each area need to know what to do if an emergency arises.

A procedure needs to be established to transmit appropriate hazard information to contractors who will be working on-site. In addition, information should be obtained from contractors regarding the hazardous materials they will bring into the facility. The individual or function responsible for obtaining and maintaining this information should be identified.

REFERENCES

"Precautionary Labeling of Hazardous Industrial Chemicals", ANSI Z129.1, (New York: American National Standards Institute, 1988).

"Chemical Hazards in the Workplace" (Washington, D.C.: American Chemical Society, 1981).

"Informing Workers of Chemical Hazards: The OSHA Hazard Communication Standard", (Washington, D.C.: American Chemical Society, 1985).

WORKPLACE HEALTH PROTECTION PROGRAM ELEMENT CHECKLIST ON HAZARD AWARENESS

		Yes	No	Don't Know	N/A
1.	Is there a written hazard awareness training program in place?	___	___	___	___
2.	Is responsibility for implementing and maintaining the hazard awareness training program clearly defined?	___	___	___	___
3.	Is the hazard awareness training activity stewarded to management?	___	___	___	___
4.	Is hazard awareness training presented to:				
	. All new employees?	___	___	___	___
	. All facility personnel?	___	___	___	___
	. Contractors?	___	___	___	___
	. Temporary assigned personnel?	___	___	___	___
5.	Are records kept of those who have received hazard awareness training?	___	___	___	___
6.	Is hazard awareness training repeated periodically for all facility personnel?	___	___	___	___
7.	Is hazard awareness training provided to potentially exposed personnel when:				
	. A new hazardous material is to be used?	___	___	___	___
	. A process change is made?	___	___	___	___
	. A person receives an assignment with potential exposure to different health hazards?	___	___	___	___
8.	Are purchase requisitions for new materials reviewed and approved before the purchase order is issued?	___	___	___	___
9.	Is a determination made to ensure that a hazardous material can be handled without excessive risk before it is purchased and used?	___	___	___	___

N/A - Not Applicable

WORKPLACE HEALTH PROTECTION PROGRAM ELEMENT CHECKLIST ON HAZARD AWARENESS

		Yes	No	Don't Know	N/A
10.	Is there a complete, up-to-date inventory of all materials in the facility?	___	___	___	___
11.	Is a complete, up-to-date set of material safety data sheets readily available?	___	___	___	___
12.	Are material safety data sheets available to personnel at work locations where exposure to each hazardous material may occur?	___	___	___	___
13.	Has an individual or group been given responsibility to ensure labeling is effective?				
14.	Are all containers properly labeled?	___	___	___	___
15.	Are personnel aware of the potential adverse health effects that the hazardous materials to which they may be exposed can cause?	___	___	___	___
16.	Are checks made regarding the effectiveness of hazardous awareness training (e.g., by questioning of employees)?	___	___	___	___
17.	Are personnel aware of the protective equipment to use when handling specific hazardous materials?	___	___	___	___
18.	Are personnel aware of what to do if an emergency arises that may result in their exposure to a hazardous material?	___	___	___	___
19.	Does the facility provide contractors hazard information (e.g., MSDSs) on materials to which contractor employees may be exposed?	___	___	___	___

N/A - Not Applicable

WORKPLACE HEALTH PROTECTION PROGRAM ELEMENT CHECKLIST ON HAZARD AWARENESS

	Yes	No	Don't Know	N/A
20. Do contractors provide the facility appropriate information (e.g., MSDSs) on materials they intend to bring on the site?	___	___	___	___
21. Are hazardous materials to be used by a contractor reviewed and approved for use before they are brought into the facility?	___	___	___	___
22. Do employees know how to obtain health hazard information on materials to which they may be exposed?	___	___	___	___
23. Are employees satisfied that health hazard awareness training is adequate?	___	___	___	___

N/A - Not Applicable

WORKPLACE HEALTH PROTECTION PROGRAM ELEMENT ON INVESTIGATIONS

WORKPLACE HEALTH PROTECTION PROGRAM ELEMENT ON INVESTIGATIONS

OBJECTIVE

To identify types of investigations to carry out for assessing the potential for adverse health effects to occur among personnel exposed to workplace health hazards or to obtain information necessary to prevent the occurrence, or recurrence, of such adverse health effects.

BACKGROUND

The purpose for investigating a real, alleged, or potential adverse health effect situation is to determine the probable or potential cause, and identify control measures for preventing additional excessive exposures. Issues to be addressed may include: whether there is adherence with established procedures and work practices; effectiveness of exposure controls; appropriateness of personal protective equipment; employee awareness of the potential hazards associated with the work; whether there is a workplace health protection program element in place which addresses the hazard awareness issue; and others.

All alleged and real excessive exposure occurrences should be investigated through a fact finding effort to determine cause rather than fault. It should be made clear to all contacts that the investigation is being conducted for the purpose of obtaining relevant information to help prevent the occurrence or recurrence of an adverse health effect. Facts should be documented accurately and a summary report prepared for management at the conclusion of the investigation.

There are two general types of investigations that can be carried out to address workplace health protection issues. One is to obtain information related to an exposure situation for which a plan is to be developed to assess exposure potential. The other is an investigation initiated by management, medical, or an accident investigating committee for the purpose of determining the cause of an adverse health effect and identifying controls necessary to prevent a recurrence. The effect, or potential effect, being investigated may involve hearing loss, the diagnosis of asbestos related disease, a high blood lead level in a welder, dermatitis, cumulative trauma disorder, or other adverse health effect resulting from exposure to a hazardous agent, using improper work procedures, working at poorly designed work stations or other cause.

The type of investigation referred to as a preliminary survey, is carried out to obtain information related to an exposure situation that may require the development and implementation of a monitoring program, when deemed necessary, to quantitatively assess exposure risk. Sources of exposure are identified, the potential for exposure is assessed, and the control measures in effect (engineering, administrative, or personal protective equipment use) are evaluated. Bulk samples may be obtained during this phase of the assessment to characterize the potential contaminants to which personnel may be exposed.

If monitoring is to be carried out, a sampling strategy is developed. It should be based on such factors as the reason for monitoring, who to monitor, when to collect samples, possible substances to sample for, the types of samples to obtain (time-weighted, full shift, short-term, ceiling, etc.), and where to obtain the samples (personal, area, source, or a combination). In addition, potential interferences are identified and decisions made as to where to have samples analyzed. Equipment needed to evaluate exposure control effectiveness is selected and other relevant information necessary to carry out the exposure assessment is obtained. For some situations, it is advisable to obtain additional samples after controls have been implemented to demonstrate that resolution of the exposure concern has been achieved.

The findings of an investigation can provide information for use in developing or enhancing the employee hazard awareness program. An ineffective effort in this training may be the reason that the excessive exposure occurred.

CONSIDERATIONS

Specific objectives should be developed for each investigation that is carried out for determining the potential for adverse health effects to occur among employees. These should be formulated and documented before the investigative effort is initiated. This will enable participants to maintain focus on the objectives over the course of the investigation.

Persons being contacted to provide information should be aware of the objectives of the investigation and encouraged to cooperate fully so that effective resolution of the exposure concern can be accomplished. It should be stressed that the investigation is not to determine fault, but rather, to identify the cause, or potential cause, of the concern being addressed.

All information related to the specific concern, or incident, should be documented. The findings and recommendations should be discussed with management and a final written report developed at the conclusion of the investigation. Distribution will depend on management's direction.

There should be sufficient follow-up of the recommendations that are made to ensure that the exposure concern has been resolved. It may be necessary to develop and implement a monitoring program or carry out a survey to demonstrate this. Biological testing may be warranted to further ensure that excessive exposures are not occurring.

REFERENCES

Munsch, M. H. and R. L. Potter, "Compliance and Projection", Patty's Industrial Hygiene and Toxicology, Volume III, Theory and Rationale of Industrial Hygiene Practice (New York: John Wiley & Sons, 1979), pp. 719-736.

"BBP Safety Management Handbook" (Waterford, CT: Bureau of Business Practices, 1986), pp. 10-1 to 10-9.

Brief, R. S., "Basic Industrial Hygiene" (Akron, OH: American Industrial Hygiene Association, 1975), pp. 14-15.

"Accident Investigations, Analysis, and Costs", Accident Prevention Manual for Industrial Operations, 7th Edition (Chicago, IL: National Safety Council, 1974), pp. 150-171.

Castleman, B. I. and G. E. Ziem, "Corporation Influence on Threshold Limit Values", American Journal of Industrial Medicine, No. 13, 1988, pp. 531-559.

WORKPLACE HEALTH PROTECTION PROGRAM ELEMENT CHECKLIST ON INVESTIGATIONS

	Yes	No	Don't Know	N/A
1. Do personnel with workplace health protection program responsibility participate in investigations that are carried out related to:				
. Occupational disease occurrences?	___	___	___	___
. Alleged excessive exposure situations?	___	___	___	___
. Adverse health effect occurrences due to hazardous material exposures?	___	___	___	___
. Hearing loss?	___	___	___	___
. Excessive ionizing radiation exposure occurrences?	___	___	___	___
. Cumulative trauma disorders?	___	___	___	___
. Heat stress disorders?	___	___	___	___
. Ventilation (local exhaust/general)?	___	___	___	___
. Exposures to hazardous materials associated with spills/releases?	___	___	___	___
. Video display unit issues?	___	___	___	___
. Employee complaints?	___	___	___	___
. Odor complaints:				
- Within the facility?	___	___	___	___
- In surrounding community?	___	___	___	___
. Indoor air quality issues?	___	___	___	___
. Interrogatories on compensation claims for occupational diseases?	___	___	___	___
. Employee complaints of unsatisfactory work environment conditions (e.g., noise, dust, vapors, etc.)?	___	___	___	___
. Other? (________________)	___	___	___	___
2. Are the objectives of each investigation clearly defined before it is initiated?	___	___	___	___
3. Is it made clear to contacts that the purpose of the investigation is to determine cause, not fault?	___	___	___	___
4. Are investigations carried out to demonstrate the effectiveness of control measures for reducing exposure to hazardous agents?	___	___	___	___

N/A - Not Applicable

WORKPLACE HEALTH PROTECTION PROGRAM ELEMENT CHECKLIST ON INVESTIGATIONS

	Yes	No	Don't Know	N/A
5. Is a written report issued presenting the findings of each investigation?	___	___	___	___
6. Where appropriate, is there follow-up to ensure measures have been implemented to reduce exposure risk and prevent a recurrence of an incident?	___	___	___	___
7. When appropriate, are biological monitoring programs implemented to demonstrate that control measures have been effective for reducing exposure to health hazards that have been investigated?	___	___	___	___

N/A - Not Applicable

WORKPLACE HEALTH PROTECTION PROGRAM ELEMENT
ON
PERMITS

WORKPLACE HEALTH PROTECTION PROGRAM ELEMENT ON PERMITS

OBJECTIVE

To identify the need to evaluate potential health hazards associated with hot work, cold or safe work, and confined space entries, and ensure that health exposure risks, in addition to safety concerns, are acceptable before a permit is issued to perform the work.

BACKGROUND

Work permit systems have been in place in many industries for a number of years. The permit provides a record to show that a particular job to be done has been checked by responsible individuals/groups, as well as what must be done to perform the work without excess risk. The procedure was initially developed to minimize the potential for fires/explosions to occur as a result of the nature of the work that was to be carried out in locations where combustible/flammable gases or vapors were present. The work permit system continues to meet that need. In addition, personnel and environmental health concerns need to be addressed.

Assessment of the health hazard potential for work permit issuance, where potential health hazards exist, requires the determination of the concentration of chemical contaminants, oxygen, and levels of physical agents to which personnel may be exposed, and of the potential for exposure to these. Based on the results, a judgement can be made regarding the measures that need to be implemented to perform the work without excessive risk to health. This may involve the application of engineering or administrative controls, the use of personal protective equipment, or a combination of these measures. The appropriate exposure control needs should be identified on the permit.

The concentration of flammable or combustible gas or vapor which may present a health risk to personnel is typically several orders of magnitude below that which may pose a fire or explosion hazard. Thus, it may be necessary to use instruments or monitoring methods which are very sensitive and specific in order to assess the exposure risk for substances that potentially present a health hazard to personnel. For example, use of a combustible gas indicator type device, typically used to determine the fire/explosion potential of an atmosphere, is generally not adequate for determining the potential inhalation hazard of a flammable/combustible contaminant.

The types of permits issued typically include those for hot work, cold work, safe work, excavations and entry (confined space). The determinations made relative to the health concerns associated with each would generally be the same for work in a given location or area, but the exposure assessment and control considerations could vary. For example, continuous monitoring may be appropriate for confined space work.

Reportedly, permits for work with asbestos-containing materials (ACM) and ionizing radiation sources are increasingly being used. The information presented in the program elements on these topics, as well as that presented in this discussion would have relevance to the issues to consider in issuing permits for ACM work or radiographic procedures.

The information developed in the exposure assessment of the task to be permitted is the basis for conveying to management and employees the measures that must be implemented to reduce the associated health exposure risks. This information is transmitted to personnel through indications/statements on the permit. The employees' supervisor has responsibility to ensure that personnel doing the work comply with these requirements, as well as with the established procedures/work practices.

CONSIDERATIONS

A work permitting system should be in place to aid in protecting personnel from injury, and equipment from damage. The system should address both safety and health concerns associated with the work to be performed. Implementing and maintaining such a procedure requires cooperation and communication between individuals requesting the permit, personnel doing the testing/evaluation for issuing the permit, those with authority to issue the permit, personnel doing the work, and supervisory personnel.

The permit should define the conditions or restrictions under which the work can proceed. The measures to be taken to control potential health hazards could be delineated on the permit.

Safe work is any work performed in an area, or the operation of any equipment in an area, which may contain flammable or hazardous substances. A cold work permit is to be issued for work where a potential physical hazard exists (flammable, explosive, rotating equipment, electrical equipment, etc.), or where there is the potential for exposure to a health hazard (toxic, corrosive, cyrogenic, carcinogenic, reproductive, noise, heat stress, etc.) and does not introduce an ignition source into the work environment which could give rise to a fire or explosion. A hot work permit is to be issued when a hazardous material may be present, the potential for fire or explosion exists in an area or inside equipment, and the work involves a source of ignition that could cause a fire or explosion. An entry permit is to be issued for those situations where entry into a confined space is required. Here, the nature of the work to be performed can be either hot, cold, or safe. If work is not started within a specified period after a permit is issued (e.g., 1 hour), the permit should be voided and another issued. In addition, a permit should be issued for only a limited time period (e.g., a shift, 4 hours, etc.). If conditions which could create a hazard change during the permitted period, the work is to be stopped, and a new permit issued after the cause for the work stoppage has been corrected.

It is essential that the individuals responsible for evaluating the health hazards associated with the permitted work use appropriate assessment methods. Monitoring equipment must be effectively maintained and in calibration. The sampling and analytical procedures employed must be those that have been demonstrated to be effective for the application. Results of measurements should be properly documented and maintained. Exposure control procedures to be implemented for performing the task should be identified according to the permit procedure by their employer. Personnel doing the permitted work must be trained to carry it out in the approved manner. The supervisor of the employees performing the task should ensure that established procedures are being followed. The inhalation and skin absorption potential, as well as the potential for ingestion of the contaminants to which personnel are exposed, should be considered in defining exposure controls.

Where warranted, periodic checks should be made to ensure that personnel performing permitted work, and the conditions of the work area, are in compliance with the specifications on the permit. This may involve going to the work location periodically to ensure that the control measures indicated on the permit have been implemented and are being complied with, and accepted procedures/work practices are being followed.

REFERENCES

"Accident Prevention Manual for Industrial Operations", 7th Edition (Chicago: National Safety Council, 1974).

"Flammable and Combustible Liquids Code Handbook", 3rd Edition (Quincy, MA: National Fire Protection Association, 1988).

Meidl, J. H., "Flammable Hazardous Materials" (Beverly Hills, CA: Glencoe Press, 1970), pp. 23-44.

WORKPLACE HEALTH PROTECTION PROGRAM ELEMENT CHECKLIST ON PERMITS

	Yes	No	Don't Know	N/A
1. Are permits issued for:				
. Hot work?	___	___	___	___
. Cold work?	___	___	___	___
. Safe work?	___	___	___	___
. Confined space entry?	___	___	___	___
. Excavation?	___	___	___	___
. Asbestos work?	___	___	___	___
. Radiography?	___	___	___	___
2. Is the permit form effective for transmitting health hazard and exposure control information to personnel performing a task?	___	___	___	___
3. Do supervisory personnel ensure that employees comply with the health hazard exposure control specifications identified on the permit form?	___	___	___	___
4. Do procedures for health hazard assessment associated with the work permit system address:				
. Chemical health hazard exposure potential:				
- Inhalation?	___	___	___	___
- Ingestion?	___	___	___	___
- Skin absorption?	___	___	___	___
- Skin/eye contact?	___	___	___	___
. Oxygen deficiency?	___	___	___	___
. Physical health hazard exposure potential:				
- Noise?	___	___	___	___
- Ionizing radiation?	___	___	___	___
- Heat stress?	___	___	___	___
- Non-ionizing radiation?	___	___	___	___
. Biological health hazard exposure potential?	___	___	___	___
. Personal hygiene?	___	___	___	___
. Ergonomic factors?	___	___	___	___
5. Are personnel who assess health hazard risk potential for permit issuance adequately trained to provide this support?	___	___	___	___

N/A - Not Applicable

WORKPLACE HEALTH PROTECTION PROGRAM ELEMENT CHECKLIST ON PERMITS

	Yes	No	Don't Know	N/A
6. Is the available instrumentation and exposure assessment method/procedure adequate for evaluating health hazard exposure potential associated with work to be permitted?	___	___	___	___
7. Are permit related monitoring results effectively documented?	___	___	___	___
8. Is responsibility identified for periodically checking to ensure that personnel adhere to the health hazard exposure control specifications delineated on permits?	___	___	___	___
9. Is a procedure in place to determine if workplace conditions have changed sufficiently to warrant voiding a permit:				
. Continuous monitoring:				
- Area?	___	___	___	___
- Personal?	___	___	___	___
. Periodic retesting?	___	___	___	___
. Comments of personnel?	___	___	___	___
. Odor perception indicating respirator leakage/breakthrough?	___	___	___	___
. Other? (________________)	___	___	___	___
10. Are records available on the maintenance and calibration of instrumentation used to assess health hazards associated with permitted work?	___	___	___	___
11. Are the methods/procedures used to assess health hazards associated with permitted work periodically reviewed to ensure they are effective for the application?	___	___	___	___

N/A - Not Applicable

WORKPLACE HEALTH PROTECTION PROGRAM ELEMENT CHECKLIST ON PERMITS

	Yes	No	Don't Know	N/A
12. In addition to the safety considerations of a job, does the permit form indicate:				
. Type of work to be performed (hot, cold, safe, entry)?	___	___	___	___
. Date permit issued?	___	___	___	___
. Description of work to be performed?	___	___	___	___
. Identification of the work location?	___	___	___	___
. Basis for voiding the permit:				
- Change in conditions?	___	___	___	___
- Work delayed beyond established time period?	___	___	___	___
- Permitted work period has elapsed?	___	___	___	___
. Person in charge of work area has checked work proposal?	___	___	___	___
. Equipment to be worked on has been:				
- Depressurized?	___	___	___	___
- Drained, purged, cleaned?	___	___	___	___
- Properly isolated?	___	___	___	___
. Sources of contamination (sewer openings, other spill/release sources) are recognized and measures have been taken to minimize the potential for a release from these?	___	___	___	___
. Specific exposure control measures (engineering, personal protective equipment, emergency equipment, etc.) to be taken?	___	___	___	___
. Potential health hazards tested for?	___	___	___	___
. Contaminant measurement results?	___	___	___	___
. Need for retesting (time, frequency)?	___	___	___	___
. Identification of person conducting the tests?	___	___	___	___
. Approval signatures?	___	___	___	___

N/A - Not Applicable

WORKPLACE HEALTH PROTECTION PROGRAM ELEMENT
ON
PROJECT REVIEW

WORKPLACE HEALTH PROTECTION PROGRAM ELEMENT ON PROJECT REVIEW

OBJECTIVE

To provide information for use by responsible groups addressing workplace health protection concerns when conducting project reviews to ensure that controls for potential health hazards associated with a process, material, or equipment used in an operation are identified and recommended for incorporation into the design.

BACKGROUND

Engineering methods for controlling the chemical, physical, biological, and ergonomic factors which can give rise to adverse workplace health effects are preferred over administrative methods or the use of personal protective equipment. It is more effective to incorporate engineering controls or other protective measures into an operation at the design stage than to undertake retrofits. When done beforehand, available options can be considered for achieving the optimum balance between cost and employee health benefit.

All projects, process and equipment changes, modified usage of chemicals or biologic agents, and plans for production increases should be reviewed by various technical functions with responsibility for workplace health protection issues. Design engineering should be apprised of existing and new/developing information related to the health hazards associated with the materials made or used in operations. Necessary changes can then be considered in the selection of exposure controls for new units or modifications to existing ones.

CONSIDERATIONS

In order to carry out an effective design review of a new facility, or modification to an existing one, it is essential to have a description, and flow diagrams of the process; a complete inventory of chemicals to be used or produced; information on the equipment to be used in the operations; wastes to be generated; and appropriate material balance information. Engineering control measures for minimizing exposure to chemical, physical and biological agents, and ergonomic factors need to be clearly delineated in equipment design specifications in order to minimize, as much as practical, potential workplace health hazards associated with their operation. The process description should provide sufficient information to enable a clear understanding of the process and the basis for the contaminant control methodologies to be incorporated. Process information should be sufficient to assure that the chemical inventory is complete. Information regarding process sampling needs; frequency of catalyst skims/change-out; process additives; waste/sludge removal needs; required personal protection; etc. should be presented. All controls should be reviewed against new or proposed regulations, and changes incorporated, where necessary, to meet the regulatory requirements. The need to install deluge showers/eyewash fountains, change rooms, shower facilities, and continuous monitors should be identified.

The chemical inventory should identify use rates for raw materials, catalysts, and additives, as well as production rates for products, by-products and chemical intermediates. The contaminants to be generated in the process, such as sludges, wastes, and potential releases to the air and water should be identified.

Process flow details should be reviewed to ensure selected control procedures are adequate. When appropriate, identify and recommend alternative methods for design consideration.

Process stream sample points should be easily accessed. Closed loop sampling methods would be appropriate when the material being sampled is a particularly toxic substance, or has a high vapor pressure.

Releases to the atmosphere from safety relief valves, vents, and stacks should be at a safe location. Vent and safety valve releases are often most effectively controlled when directed to a flare.

Chemicals are often added to the process through mixing drums or hoppers. These may need to be provided with local exhaust ventilation, depending on the hazardous properties of the material to be added, the likelihood that the operator will be exposed to the material, as well as the frequency and duration for performing the task.

When practical, noise criteria should be specified for equipment. This information should be provided to the vendor as a part of the purchase requisition. Noise attenuators or enclosures should be considered for motors, pumps, and compressors. Noise emissions from pressure regulators, and recycle/control valves may have to be reduced. Piping carrying fluids (gases or liquids) at high velocity, or piping upstream and downstream from noisy valves/regulators, should also be considered for noise attenuation. When noise control is not practical, shelters should be installed for personnel to provide isolation from the noise.

Ventilation system design parameters (capture/control velocity, exhaust rate, transport velocity, duct size, construction materials, cleanout ports, etc.) should be identified so they can be considered relative to properties of the contaminant being controlled. Where appropriate, provisions for adequate make-up air must be considered. Make-up air should be supplied from a location where contaminant re-entry or entrainment will not occur. System performance should be checked during commissioning to ensure design specifications are met.

The types of seals specified for pumps and compressors should be based on their effectiveness for minimizing leakage when in service. Double mechanical seals, oil seals, or labyrinth seals are often used to reduce leakage in operations where single seals were previously used.

Spills of toxic materials should be contained. Toe walls and surface drain needs should be evaluated relative to the properties of the materials that may be released. Cleanout connections should be provided to facilitate equipment cleaning (pumps, strainers, exchangers, etc.) and thereby eliminate draining contents to grade. Flushings from equipment should be directed into a closed system to minimize release to the air and facilities to accomplish this should be incorporated into the design. For some equipment, permanent closed systems may be necessary.

Appropriate insulation should be provided on hot process equipment. This will minimize the likelihood for burns, as well as reduce heat stress on personnel working in the area. Shielding should be provided, where appropriate, to reduce radiant heat. When necessary, exhaust ventilation should be installed to remove heat generated by the equipment/process. Conditioned air should be provided, where appropriate.

Where there is the possibility for release of acutely acting, or highly toxic contaminants from the process or equipment, it may be advisable to provide continuous monitors to detect releases and alert personnel of the potential hazard. This equipment is typically installed near potential sources of release, within a make-up air system, or in areas where personnel normally work. Sensor and alarm locations should be selected by industrial hygiene, environmental, safety, and others.

Valves, flanges, pump/compressor seals, and other equipment can be significant sources of fugitive emissions. Many require frequent maintenance and replacement of packing to eliminate or reduce leaks. Alternatives to asbestos packing material, such as graphite types, are available. The graphite packing, in conjunction with lantern rings, and an extended stuffing box with a purge system should be considered for carcinogenic, mutagenic, or high toxicity gases/vapors.

REFERENCES

Lipton, S. and J. R. Lynch, "Health Hazard Control in the Chemical Process Industry" (New York: John Wiley & Sons, Inc., 1987).

Cralley, L. V. and L. J. Cralley (eds.), "In-Plant Practices for Job Related Health Hazard Control - Production Processes", Vol. 1 (New York: John Wiley & Sons, Inc., 1989).

Cralley, L. V. and L. J. Cralley (eds.), "In-Plant Practices for Job Related Health Hazard Control - Engineering Aspects", Vol. 2 (New York: John Wiley & Sons, Inc., 1989).

Soule, R. D., "Industrial Hygiene Engineering Controls", Patty's Industrial Hygiene and Toxicology - General Principles", Vol. 1, 3rd Edition (New York: John Wiley & Sons, Inc., 1978), pp. 771-823.

"Industrial Ventilation - A Manual of Recommended Practice", Latest Edition (Lansing, MI: American Conference of Governmental Industrial Hygienists).

Harris, C. M., (ed.), "Handbook of Noise Control", 2nd Edition (New York: McGraw-Hill Book Company, Inc., 1979).

"Inspection for Accident Prevention in Refineries", API Publication 2002 (Washington, D. C.: American Petroleum Institute, 1984).

Shapiro, J., "Radiation Protection", 3rd Edition (Cambridge, MA: Harvard University Press, 1990).

Cralley, L. V. and L. J. Cralley (eds.), "Industrial Hygiene Aspects of Plant Operations", Vol. 3, Engineering Considerations in Equipment Selection, Layout, and Building Design (New York: Macmillan Publishing Co., 1985).

WORKPLACE HEALTH PROTECTION PROGRAM ELEMENT CHECKLIST ON PROJECT REVIEW

		Yes	No	Don't Know	N/A
1.	Is the industrial hygiene function involved in site selection discussions?	___	___	___	___
2.	Are nearby process operations a consideration in site selection relative to the potential for their releasing hazardous substances?	___	___	___	___
3.	Is adequate information provided to enable an effective design review:				
	. Narrative?	___	___	___	___
	. Process description?	___	___	___	___
	. Equipment information?	___	___	___	___
	. Inventory of chemicals?	___	___	___	___
	. Basis for controls?	___	___	___	___
	. Location of facility?	___	___	___	___
	. Other operations in the vicinity?	___	___	___	___
4.	Do groups with responsibility for workplace health protection review plans and specifications of new operations or changes to existing ones?	___	___	___	___
5.	Do groups with workplace health protection program responsibilities sign off to indicate that a review of a design has been carried out?	___	___	___	___
6.	Is there effective communication between the design group and others who review project designs (industrial hygiene, safety, environmental, etc.)?	___	___	___	___
7.	Is there an effective method for transmitting workplace health protection type recommendations on project design to personnel responsible for implementing/ effecting design changes?	___	___	___	___

N/A - Not Applicable

WORKPLACE HEALTH PROTECTION PROGRAM ELEMENT CHECKLIST ON PROJECT REVIEW

		Yes	No	Don't Know	N/A
8.	Is there follow-up by responsible functions to determine if recommendations regarding design considerations were implemented?	___	___	___	___
9.	Is there follow-up by the industrial hygiene function to determine (after process is operational) whether exposure controls incorporated in the design were effective as determined by:				
	. Area monitoring?	___	___	___	___
	. Personal monitoring?	___	___	___	___
	. Source monitoring?	___	___	___	___
	. Ventilation system performance evaluation?	___	___	___	___
10.	Are the findings of follow-up studies documented?	___	___	___	___
11.	Does the design group meet periodically with various technical functions (industrial hygiene, safety, environmental, etc.) to discuss concerns, new issues, project status, changes, etc.?	___	___	___	___
12.	Are groups with workplace health protection program responsibility contacted with regard to specifying control measures to eliminate, where feasible, the use of personal protective equipment?	___	___	___	___
13.	Are processes which handle acutely toxic materials (e.g., H_2S, SO_2, Cl_2, NH_3, etc.) provided with a continuous air monitoring system that will warn of a significant release?	___	___	___	___

N/A - Not Applicable

WORKPLACE HEALTH PROTECTION PROGRAM ELEMENT CHECKLIST ON PROJECT REVIEW

		Yes	No	Don't Know	N/A
14.	Are critical health hazard control devices provided with an alarm which is activated in case of failure?	___	___	___	___
15.	Are deluge shower/eyewash fountain facilities adequate relative to location, design parameters, and needs?	___	___	___	___
16.	Are additive addition locations provided with:				
	. An adequate working platform?	___	___	___	___
	. Local exhaust ventilation, if appropriate?	___	___	___	___
	. Adequate lighting?	___	___	___	___
17.	Are flanged connections minimized, when warranted?	___	___	___	___
18.	Are protective equipment stations and supplied breathing air connections effectively located and identified throughout the facility?	___	___	___	___
19.	Are shelters/enclosures provided in the work area for protection from:				
	. Noise?	___	___	___	___
	. Heat?	___	___	___	___
	. Cold?	___	___	___	___
20.	Are noise specifications a part of equipment purchase specifications?	___	___	___	___
21.	Are alternatives provided to design personnel for equipment which warrants noise control?	___	___	___	___
22.	Are noise controls specified for equipment, such as:				
	. Enclosure or isolation of turbo-generators/motors?	___	___	___	___

N/A - Not Applicable

WORKPLACE HEALTH PROTECTION PROGRAM ELEMENT CHECKLIST ON PROJECT REVIEW

	Yes	No	Don't Know	N/A
. Noise attenuators for pump motors?	___	___	___	___
. Sound insulation for noise sources:				
- Lines?	___	___	___	___
- Valves?	___	___	___	___
- Pressure regulators?	___	___	___	___
. Forced air draft for furnaces/boilers?	___	___	___	___
. Sound barriers between frequently maintained equipment?	___	___	___	___
23. Are heat sources provided with shielding or insulation to reduce radiant heat emissions?	___	___	___	___
24. Were ergonomic factors considered in equipment selection, design, siting, etc.?	___	___	___	___
25. Where appropriate, was a dedicated breathing air system considered for the work area?	___	___	___	___
26. Are hose connections for the supplied breathing air system incompatible with the facility process/instrument air, nitrogen, steam, or other fittings?	___	___	___	___
27. Are double mechanical seals provided on pumps when necessary to minimize leakage?	___	___	___	___
28. Do vacuum pump seal oil vents discharge to unoccupied locations?	___	___	___	___
29. Do compressor seal oil vents discharge to unoccupied locations?	___	___	___	___
30. Are relief valves installed so that they release at a safe location?	___	___	___	___

N/A - Not Applicable

WORKPLACE HEALTH PROTECTION PROGRAM ELEMENT CHECKLIST ON PROJECT REVIEW

		Yes	No	Don't Know	N/A
31.	Is equipment (strainers, pumps, etc.) designed in parallel so that one may be removed from service without disrupting the process?	___	___	___	___
32.	Have dead ends been eliminated in piping design?	___	___	___	___
33.	Can equipment be drained effectively into a closed system?	___	___	___	___
34.	Is there adequate provision (e.g., fittings) for flushing/cleaning/ purging equipment?	___	___	___	___
35.	Are process sample points easily accessible?	___	___	___	___
36.	Were the number of process sample points minimized?	___	___	___	___
37.	Are process sample points for volatile and highly toxic materials a closed loop type?	___	___	___	___
38.	Where appropriate, are sample points provided with a hood or splash guard?	___	___	___	___
39.	Are sample points provided with a means to dispose of sample line flushings?	___	___	___	___
40.	Is there adequate lighting at sample points?	___	___	___	___
41.	Are general lighting design levels adequate?	___	___	___	___
42.	Is emergency lighting provided?	___	___	___	___

N/A - Not Applicable

WORKPLACE HEALTH PROTECTION PROGRAM ELEMENT CHECKLIST ON PROJECT REVIEW

	Yes	No	Don't Know	N/A
43. Are continuous air monitors:				
. Selected to minimize interference effects?	___	___	___	___
. Effectively sited for the purpose intended?	___	___	___	___
. Provided with an effective alarm system:				
- Audible (near sensor and in control room)?	___	___	___	___
- Visible (near sensor and in control room)?	___	___	___	___
- Readily seen and heard in area where located?	___	___	___	___
44. Are continuous air monitors considered critical equipment?	___	___	___	___
45. Are air inlets to buildings/ facilities situated away from hazardous material discharge points in order to prevent re-entrainment problems?	___	___	___	___
46. Is there adequate provision for supplying tempered make-up air to areas provided with mechanical exhaust ventilation?	___	___	___	___
47. Are laboratory hoods:				
. Properly designed (i.e., airfoil design)?	___	___	___	___
. Provided with adequate exhaust?	___	___	___	___
. Located outside high traffic areas?	___	___	___	___
. Positioned so that supply air to the laboratory does not create disturbance at the hood face?	___	___	___	___
48. Are laboratory areas in multi-use buildings at slight negative pressure with respect to surroundings (e.g., offices, corridors, etc.)?	___	___	___	___

N/A - Not Applicable

WORKPLACE HEALTH PROTECTION PROGRAM ELEMENT CHECKLIST
ON
PROJECT REVIEW

	Yes	No	Don't Know	N/A
49. Is exhaust ductwork properly sized to provide adequate transport velocity?	___	___	___	___
50. Are clean out ports provided in ductwork?	___	___	___	___
51. Are appropriate openings provided in ductwork/stack for ventilation system testing?	___	___	___	___
52. Is fan construction and ductwork the proper type for the material being exhausted?	___	___	___	___
53. Are concentric or offset type exhaust stacks used?	___	___	___	___
54. Are exhaust rates adequate at:				
. Laboratory hoods?	___	___	___	___
. Welding hoods?	___	___	___	___
. Blending tanks?	___	___	___	___
. Grease kettles?	___	___	___	___
. Additive tanks?	___	___	___	___
. Other (________________)?	___	___	___	___
55. Are design reviews carried out:				
. At the process description stage?	___	___	___	___
. Of the site layout before construction?	___	___	___	___
. Prior to AFE (Authorization for Expenditure) approval?	___	___	___	___
. At preconstruction or model stage?	___	___	___	___
. At interim construction stage (50%) completion?	___	___	___	___
. Pre-start-up?	___	___	___	___
. Post start-up?	___	___	___	___

N/A - Not Applicable

WORKPLACE HEALTH PROTECTION PROGRAM ELEMENT CHECKLIST ON PROJECT REVIEW

	Yes	No	Don't Know	N/A
56. Do groups with workplace health protection program responsibility consider the following in design reviews:				
. Inventory of materials to be used?				
. Potential chemical contaminants or physical/biologic agents to which personnel may be exposed?				
. Locations/operations where exposures may occur?				
. Illumination?				
. Ventilation?				
. Heat stress?				
. Pump/compressor seals and material selection?				
. Vent (sewer, compressor seal, oil drum, safety relief valves, etc.) location?				
. Paint specifications?				
. Applications where double mechanical seals are appropriate for control?				
. Insulation composition?				
. Gasket composition?				
. Noise:				
- Sources?				
- Controls?				
. Ergonomic factors?				
. Siting of emergency equipment (e.g., self-contained breathing air units, deluge showers/eyewash fountains, etc.)?				
. Solids handling facilities (catalyst, filter aid, additives, etc.)?				
. Laboratory design?				
. Selection and locating of continuous monitors?				
. Access/egress for confined space work locations (process vessels, tanks, etc.)?				
. Equipment draining and flushing considerations?				

N/A - Not Applicable

WORKPLACE HEALTH PROTECTION PROGRAM ELEMENT CHECKLIST ON PROJECT REVIEW

	Yes	No	Don't Know	N/A
. Location and orientation of radiation sources?	___	___	___	___
. Process sampling points:				
- Location?	___	___	___	___
- Design?	___	___	___	___
. Curbing (toe wall) around equipment?	___	___	___	___
. Applications for use of process analyzers rather than collection of process samples?	___	___	___	___
. Are employees identified before unit/process start-up for inclusion in hazard awareness training and medical surveillance programs (e.g., respirator, benzene, etc.)?	___	___	___	___
. Other (________________)?	___	___	___	___

N/A - Not Applicable

WORKPLACE HEALTH PROTECTION PROGRAM ELEMENT
ON
PURCHASING

WORKPLACE HEALTH PROTECTION PROGRAM ELEMENT ON PURCHASING

OBJECTIVE

To ensure that the potential hazards associated with the use of a material are known before purchase, and appropriate exposure controls are implemented prior to receipt and/or use.

BACKGROUND

The total removal of all potentially harmful agents from a facility would be a method for preventing adverse health effects that may result from their use. This, however, is not practical. It is possible, in some situations, to prevent their introduction by substituting suitable, less hazardous alternatives. This too, is often impractical or impossible. By instituting appropriate control procedures before purchase and use, it is often possible to reduce an exposure hazard to an acceptable level and thereby enable the use of a hazardous substance without excessive risk.

When the hazards of a material and its intended use are known beforehand, appropriate measures can be taken to minimize, or reduce the potential for adverse health effects to arise. Thus, it is essential that the potential hazards of a material be recognized before it is introduced into the workplace.

In many jurisdictions, the purchase and use of sources of ionizing radiation require approval by the Radiation Safety Officer (RSO) and a license and/or registration must be obtained before a purchase order is issued. The radiation source shipment should be checked when received to ensure that the external radiation level is acceptable, the source is intact, and it is properly labeled. A similar practice could be implemented for other hazardous substances.

CONSIDERATIONS

It is advisable to determine the properties of a material that will be used in a facility, as well as the method of use (where and how, how much, who, how often, etc.) before its purchase and receipt so that appropriate control measures can be developed and implemented for reducing potential health hazards. A system should be in place which enables a review of the purchase requisitions for new materials by various functions (industrial hygiene, safety, environmental) before issuing the purchase order. A purchasing policy should be in place to ensure that health hazards are considered when reviewing materials for purchase so that less hazardous products can be selected if available. In addition, controls should be established which prevent the introduction of materials by any method other than through the normal requisition/purchasing procedure.

Generally, a judgment of the potential hazards and exposure control needs can be made based on information provided by the supplier (e.g., material safety data sheet for the product or the vendor sound level report for equipment). Purchasing, or the user group, should obtain such information as early as possible and make it available to the reviewers well in advance of the time of need or purchase of the product so that decisions can be made with respect to exposure control requirements.

REFERENCES

"The Industrial Environment - Its Evaluation and Control", (Washington, D. C.: United States Department of Health, Education, and Welfare, 1973).

"Fundamentals of Industrial Hygiene", 2nd Edition (Chicago, IL: National Safety Council, 1985), pp. 810-811.

WORKPLACE HEALTH PROTECTION PROGRAM ELEMENT CHECKLIST ON PURCHASING

	Yes	No	Don't Know	N/A
1. Is there a procedure in place which requires purchase requisition review and approval (for new materials) by various groups (industrial hygiene, safety, environmental) before purchase of a hazardous substance?	___	___	___	___
2. Is there a complete chemical inventory of materials used/made/stored on site?	___	___	___	___
3. Is there follow-up by industrial hygiene on hazardous material purchases to ensure that the material is being used as was indicated prior to its purchase?	___	___	___	___
4. Is the chemical inventory reviewed at least annually?	___	___	___	___
5. Are copies of MSDSs that are received with purchased products forwarded to users and designated individuals/groups?	___	___	___	___
6. Are personnel assigned industrial hygiene function responsibility provided a copy of the MSDS?	___	___	___	___
7. Are requisitions for purchase of equipment containing radioactive materials reviewed and approved by the Radiation Safety Officer (RSO) before the purchase is made?	___	___	___	___
8. Is a radiation survey conducted when radioactive sources are received:				
. External radiation level determination?	___	___	___	___
. Wipe test of source?	___	___	___	___

N/A - Not Applicable

WORKPLACE HEALTH PROTECTION PROGRAM ELEMENT CHECKLIST ON PURCHASING

		Yes	No	Don't Know	N/A
9.	Do appropriate functional groups (industrial hygiene, safety, environmental, etc.) ensure purchase orders reflect appropriate specifications for visual display equipment, radiation sources, noise producing equipment, etc.?	___	___	___	___

N/A - Not Applicable

WORKPLACE HEALTH PROTECTION PROGRAM ELEMENT
ON
RECORDKEEPING

WORKPLACE HEALTH PROTECTION PROGRAM ELEMENT ON RECORDKEEPING

OBJECTIVE

To provide information on the types of workplace health protection records and documentation that should be developed and maintained.

BACKGROUND

There is a need for effective documentation of workplace health protection program activities. This is, in part, the result of regulatory requirements and worker concern. However, there are other reasons for developing and maintaining detailed information of program activities. These include the following:

- Supportive documentation on monitoring results that are provided employees and management.
- To demonstrate the effectiveness of, or need for exposure controls.
- As a basis for the selection of appropriate respiratory protection for use by employees.
- As an information base in hazard communication training.
- For epidemiologic studies.
- Identifying trends.
- For medical reference.
- Determining the potential impact of future regulations.
- As a basis for planning.
- In litigation.

Records must be complete, orderly, and readily retrievable. They should include, but are not limited to: material safety data sheets; monitoring data; equipment maintenance and calibration information; ionizing radiation source and exposure records; employee training records; etc. All information should be reviewed before filing/data entry to ensure it is accurate.

CONSIDERATIONS

Information on workplace health protection activities should be effectively reviewed and filed for reference, as needed, and in keeping with applicable regulatory requirements and company policy. The general classes of information that should be developed and retained includes:

- Copies of corporate policies and guidelines relating to workplace health protection issues.
- An inventory of all materials that are on site. This includes purchased products, raw materials, process streams, additives, intermediates, products, wastes, and others. It is advantageous if these are identified with respect to their location within the facility.
- An up-to-date compilation of material safety data sheets (MSDSs).
- Representative, personal, and other type monitoring results.
- Records of employee training as required by company guidelines and/or regulation.
- Reports of survey findings.
- Reports of industrial hygiene investigations.
- Reports of workplace health protection program reviews (audits) by internal and external groups.
- A copy of the annual workplace health protection plan.
- Copies of radiation records (licenses, registrations, location of sources, wipe test results, survey results, personal exposure data, etc.).
- Results of the findings of follow-up assessments of previously made recommendations.
- Copies of written workplace health protection programs as required by regulation or internal guidelines.
- Copies of inspections/investigations/surveys carried out by regulatory agency personnel.
- Copies of qualitative exposure assessment results.

- Instrument calibration and maintenance records.
- Biological monitoring results as provided by the Medical Department for follow-up/investigation, etc.
- Copies of information provided to contractors.
- Listing of tankage that has been in leaded fuel service.
- Listing of equipment containing PCBs.
- Locations where asbestos-containing material is present.
- Locations where NORM is present.
- Listing of equipment that has been painted with lead-based, or chromate-based paint.

WORKPLACE HEALTH PROTECTION PROGRAM ELEMENT CHECKLIST ON RECORDKEEPING

		Yes	No	Don't Know	N/A
1.	Is there a complete inventory of all potentially hazardous materials that are on site?	___	___	___	___
2.	Are MSDSs available on all materials that are on site?	___	___	___	___
3.	Are MSDSs effectively filed so they are readily retrievable?	___	___	___	___
4.	Are MSDSs up-to-date?	___	___	___	___
5.	Are records (regarding workplace health protection issues) available with respect to employee training:				
	. Hazard communications?	___	___	___	___
	. Hazardous waste operations?	___	___	___	___
	. Hearing conservation?	___	___	___	___
	. Hearing protection?	___	___	___	___
	. Respiratory protection?	___	___	___	___
	. Where applicable, for:				
	- Asbestos?	___	___	___	___
	- Benzene?	___	___	___	___
	- Confined space activities?	___	___	___	___
	- Lead?	___	___	___	___
	- Ionizing radiation?	___	___	___	___
	- Other? (__________________)	___	___	___	___
6.	Are written programs available on specific workplace health protection issues (e.g., respiratory protection, hearing conservation, hazard communications, etc.)?	___	___	___	___
7.	Are representative and/or personal monitoring results available for personnel exposed to:				
	. Asbestos?	___	___	___	___
	. Benzene?	___	___	___	___
	. Formaldehyde?	___	___	___	___
	. Ionizing radiation:				
	- Pocket dosimeter?	___	___	___	___
	- Film badge or TLD?	___	___	___	___
	. Lead?	___	___	___	___
	. Noise?	___	___	___	___
	. Silica?	___	___	___	___
	. Other? (__________________)	___	___	___	___

N/A - Not Applicable

WORKPLACE HEALTH PROTECTION PROGRAM ELEMENT CHECKLIST
ON
RECORDKEEPING

	Yes	No	Don't Know	N/A
8. Are reports of comprehensive industrial hygiene type surveys of units/areas/facilities available?	___	___	___	___
9. Are reports of findings available on surveys for:				
. Asbestos?	___	___	___	___
. Benzene?	___	___	___	___
. Lighting?	___	___	___	___
. Noise?	___	___	___	___
. Radiation?	___	___	___	___
. Fugitive emissions?	___	___	___	___
. Community noise?	___	___	___	___
. Other? (____________________)	___	___	___	___
10. Are the findings of workplace health protection program reviews (audits) available?	___	___	___	___
11. Is a copy of the annual workplace health protection program plan available?	___	___	___	___
12. Are the results of workplace health related investigations documented and available?	___	___	___	___
13. Are the results of follow-up studies/ assessments to determine the effectiveness of control measures documented and available?	___	___	___	___
14. Are copies of written compliance programs (as required by regulation and/or internal policy) available?	___	___	___	___
15. Are copies of qualitative exposure assessments available?	___	___	___	___

N/A - Not Applicable

WORKPLACE HEALTH PROTECTION PROGRAM ELEMENT CHECKLIST ON RECORDKEEPING

		Yes	No	Don't Know	N/A
16.	Are medical surveillance results (as provided by medical - e.g., audiometric test results, blood/urine lead levels, etc.) available for review and follow-up, when appropriate?	___	___	___	___
17.	Are in-house instrument calibration data documented in a permanent type binder?	___	___	___	___
18.	Are copies of operating manuals available for all instruments?	___	___	___	___
19.	Are instrument calibration data filed with operating manuals?	___	___	___	___
20.	Are copies of the facility's emergency plans available?	___	___	___	___
21.	Are copies of completed field sampling data sheets available?	___	___	___	___
22.	Are results of ventilation system surveys available for:				
	. Laboratory?	___	___	___	___
	. Mechanical shops?	___	___	___	___
	. Process sample locations?	___	___	___	___
	. Solids transfer hoppers?	___	___	___	___
	. Carpenter shop?	___	___	___	___
	. Paint spray booths?	___	___	___	___
	. Abrasive cleaning hoods/booths?	___	___	___	___
	. Plate and frame filters?	___	___	___	___
	. Other? (____________________)	___	___	___	___
23.	Are there written procedures available for contaminants that are monitored?	___	___	___	___
24.	Are records available of tank service relative to their having contained leaded gasoline or leaded aviation gasoline?	___	___	___	___

N/A - Not Applicable

WORKPLACE HEALTH PROTECTION PROGRAM ELEMENT CHECKLIST ON RECORDKEEPING

	Yes	No	Don't Know	N/A
25. Are records available for equipment/area which:				
. Has asbestos-containing material on/in it?	___	___	___	___
. Contains PCBs?	___	___	___	___
. Has NORM present?	___	___	___	___
. Has been painted with lead-based paint?	___	___	___	___
. Has been painted with a chromate-based paint?	___	___	___	___

N/A - Not Applicable

WORKPLACE HEALTH PROTECTION PROGRAM ELEMENT
ON
REGULATIONS

WORKPLACE HEALTH PROTECTION PROGRAM ELEMENT ON REGULATIONS

OBJECTIVE

To provide information relative to the role the industrial hygiene function could perform regarding regulations which impact the facility's workplace health protection program.

BACKGROUND

Legislation has been enacted in a number of countries which enables the development of regulations/standards related to workplace health protection issues. Their impact can vary from little to very great. It may be necessary to train additional personnel to assist in carrying out a plan to achieve compliance with the regulatory requirements. For some regulations, it may be necessary to develop a monitoring program, including sampling/analytical method selection, obtaining additional monitoring equipment, purchase training guides and personal protective equipment, and to take other measures, as necessary, to achieve compliance. It is essential that the functional groups having responsibility for implementing measures to respond to the requirements of workplace health protection type regulations be aware of their specific details. Management needs to be kept aware of the regulatory requirements, the plans for achieving compliance with them, as well as the compliance status.

CONSIDERATIONS

Management should identify an individual or group to review proposed and enacted regulations/standards, identify compliance needs and potential compliance problems, and provide direction for achieving compliance. This person or group should keep management apprised of the impact of specific regulations on the facility's operations and of the plans to achieve compliance.

It is essential that responsible individuals/groups be aware of the specific details of the regulations which impact the workplace health protection program. Copies of all such regulations should be readily available and reviewed periodically to ensure all requirements are being complied with. If deficiencies are noted, this information should be communicated to management and an action plan developed to establish, or re-establish compliance.

The industrial hygiene function should be aware of the requirements of workplace health regulations under development that could affect plant operations. These should be reviewed in detail as early in their development as possible so that a determination can be made regarding their potential impact on operations and resources. An early decision should be made as to whether an assessment should be carried out to identify possible compliance problems. Based on the information obtained, a plan could be developed for achieving compliance with the specific requirements of the regulation. Management should be informed if difficulties are expected in achieving compliance.

Detailed reviews could be carried out periodically by an off-site group that has been assigned the task of determining the status of compliance with those workplace health protection regulations with which the facility must comply. The site industrial hygienist, or facility personnel with workplace health protection program responsibilities, must be available to respond to audit team members questions during such reviews.

WORKPLACE HEALTH PROTECTION PROGRAM ELEMENT CHECKLIST ON REGULATIONS

		Yes	No	Don't Know	N/A
1.	Has an individual or group been designated to keep abreast of legislative developments and workplace health regulations in order to keep management informed of their potential impact?	___	___	___	___
2.	Is there a means for promptly obtaining copies of proposed and newly issued regulations that impact the facility's workplace health protection program?	___	___	___	___
3.	Are copies of all applicable workplace health protection regulations readily available to facility personnel?	___	___	___	___
4.	Does the industrial hygiene function provide input to management regarding the impact of workplace health protection regulations?	___	___	___	___
5.	Do personnel/groups with responsibility for specific aspects of the workplace health protection program develop plans for timely response to regulatory requirements?	___	___	___	___
6.	Are plans that have been developed to respond to workplace health protection regulations reviewed by management before implementation?	___	___	___	___
7.	Are potential regulatory compliance problems promptly communicated to management so that an appropriate action plan can be developed to achieve compliance?	___	___	___	___
8.	Is there an effective sampling/analytical method available to carry out the monitoring requirements of standards that impact operations?	___	___	___	___

N/A - Not Applicable

WORKPLACE HEALTH PROTECTION PROGRAM ELEMENT CHECKLIST ON REGULATIONS

		Yes	No	Don't Know	N/A
9.	Is a preliminary monitoring program typically conducted to identify needs/problems that may be encountered in complying with an impending regulation with a monitoring component?	___	___	___	___
10.	Are periodic reviews carried out to determine compliance status with all sections of each workplace health protection regulation/standard which impacts operations?	___	___	___	___
11.	Is management promptly informed of noncompliant situations?	___	___	___	___
12.	Where required by regulation or company guidelines, are employees informed of personal and/or representative monitoring results?	___	___	___	___
13.	Where required by regulation, are excessively exposed employees informed of the control measures being implemented to reduce their exposure to a contaminant?	___	___	___	___
14.	Are employees provided hazard awareness training regarding the chemical, physical, and biological agents to which they may be exposed in their employment?	___	___	___	___
15.	Is hazard awareness training repeated periodically?	___	___	___	___
16.	Is there an information exchange capability which enables one facility to benefit from the experience of another relative to effecting and maintaining regulatory compliance with workplace health protection issues?	___	___	___	___

N/A - Not Applicable

5. PERSONAL PROTECTIVE EQUIPMENT

WORKPLACE HEALTH PROTECTION PROGRAM ELEMENT
ON
DELUGE SHOWER/EYEWASH FOUNTAIN

WORKPLACE HEALTH PROTECTION PROGRAM ELEMENT ON DELUGE SHOWER/EYEWASH FOUNTAIN

OBJECTIVE

To provide information on the installation, use, and maintenance of deluge showers/eyewash fountains.

BACKGROUND

There are a large number of materials routinely used in industrial operations that can produce skin or eye damage or be absorbed through the skin. Water is generally the best, and often the most readily available material for the removal of chemicals from the skin or eyes. For substances that can be absorbed through the skin, such as phenol or hydrazine, prompt removal is critical. For materials that can produce serious eye damage within a very short time, such as sodium hydroxide and monoethanolamine, prompt flushing of the eyes with potable water is essential to prevent serious effects. Deluge showers/eyewash fountains are available for installation in a facility to accomplish this.

It is essential that an acceptable water supply be readily available if an emergency arises. However, an adequate supply of potable water is often not provided. For example, the following have been observed:

- Discharge rate and direction of flow from an eyewash fountain was inadequate.
- Discharge of hot water occurred from a deluge shower/eyewash fountain.
- Discharge of dirty (rusty) water resulted when a deluge shower/eyewash fountain was activated.
- Excessive water pressure and flow of non-aerated water resulted when an eyewash fountain was activated.
- Accumulation of ice under a deluge shower was observed due to water leakage at the shower head.
- Delays in water discharge of up to a minute or more when the shower was activated.
- Water supply to remote locations was turned off.
- Materials stored in area, thereby blocking access to the emergency equipment.
- No periodic operational checks made of deluge shower/eyewash fountain.
- Equipment not installed at locations where needed.
- Personnel not trained in locating, activating, or using this emergency equipment.
- Personnel not trained to assist others in an emergency situation where this equipment was to be used.
- Potable water not provided at the deluge shower/eyewash fountain.
- No floor drain provided.

CONSIDERATIONS

Design specifications for this type emergency equipment (deluge showers/eyewash fountains) have been developed by the American National Standards Institute (ANSI Z358.1-1981). This document, as well as local regulations, or national standards should be referred to if a deluge shower/eyewash fountain is to be installed. Units commercially available in the U.S.A. are designed to meet appropriate standards and regulatory requirements.

A clean, plentiful supply of potable water is required for effective removal of a potentially harmful agent that has contacted the skin or eyes. The shower should discharge 50 gallons per minute and the eye fountain 2.5 gallons per minute at less than 25 pounds per square inch pressure (psi) and be at an acceptable temperature (60-95 F). Use of cold water can result in shock or hypothermia. If the water is too hot, thermal injury can result. In addition, improper water temperature can result in ineffective use of a deluge shower/eyewash fountain, since personnel may not remain in the stream for sufficient time to remove all the contaminant.

The equipment should be checked periodically (by process and maintenance personnel) to assure it is operating properly. The frequency of these checks depends on the properties of the chemicals of concern, the frequency of potential exposure, exposure duration, and the likelihood for contact to occur. A daily check would be appropriate when there is frequent potential exposure to materials such as sodium hydroxide. Operational checks should be made at least weekly at other locations with a record maintained of the findings.

Piping to the shower should be insulated to prevent heat-up or freezing of the water. If the equipment is exposed in the cold weather, a method for heating the water should be provided. In addition, it is necessary to assure that the water pressure is adequate to supply sufficient flow to remote locations.

The emergency deluge shower/eyewash fountain should be located as near as possible to the potential exposure location. A distance of no more than 25 feet is preferred. In no case should the distance be more than 50 feet with travel time to the installation not more than 10 seconds.

The equipment should be provided with stay-open valves and be adequately lighted. It should have a distinguishing light and sign posted to indicate its location, when not readily apparent.

It would be advisable to provide the same type and color for this equipment throughout the facility. The height of the pull chain for the shower, the shower head, and the eyewash fountain should be standardized at about 70, 90, and 40 inches respectively.

It is necessary to ensure that drainage of water from the deluge shower/eyewash fountain piping and the pad underneath is complete. This reduces the potential for slipping and build-up of ice in cold weather.

Personnel should be trained in the use of the equipment, the procedures for assisting others in using it when the need occurs, and of the procedures for checking its operation. Employees should be encouraged to check equipment operation during each shift.

It is advisable to install an audible alarm at each deluge shower/eyewash fountain which will alarm there, as well as in an occupied location (e.g., control room). The alarms are to be activated when the equipment is used.

REFERENCES

"Accident Prevention Manual for Industrial Operations", 8th Edition, Volume I - Administration and Programs (Chicago, IL: National Safety Council, 1980).

Stein, M.I., "Safety Showers Are in From the Cold", Safety and Health Journal, June 1989.

ANSI Standard No. Z358.1-1981, "Emergency Eyewash and Shower Equipment" (New York: American National Standards Institute, 1981).

WORKPLACE HEALTH PROTECTION PROGRAM ELEMENT CHECKLIST
ON
DELUGE SHOWER/EYEWASH FOUNTAIN

		Yes	No	Don't Know	N/A
1.	Are hazardous materials with properties that could cause skin or eye injury handled in the facility?	___	___	___	___
2.	Is emergency deluge shower/eyewash fountain equipment located nearby to where exposure to these substances could occur?	___	___	___	___
3.	Have personnel been trained in the use of deluge showers/eyewash fountains?	___	___	___	___
4.	Are personnel trained to help others who have a need to use a deluge shower/eyewash fountain?	___	___	___	___
5.	If contact lens use is permitted in areas/jobs where eye contact with a hazardous material may occur, have personnel been trained to provide appropriate assistance if an emergency arises?	___	___	___	___
6.	Are deluge showers/eyewash fountains:				
	. Within 50 feet of the point where contact with hazardous substances is likely to occur?	___	___	___	___
	. Clearly seen from the point where contact with skin/eye hazards could occur?	___	___	___	___
	. Provided with potable water?	___	___	___	___
	. Adequately illuminated?	___	___	___	___
	. Provided with a distinctive sign and color to identify location?	___	___	___	___
	. Situated away from electrical equipment?	___	___	___	___
	. Equipped with a local alarm?	___	___	___	___
	. Equipped with a remote alarm in an occupied area?	___	___	___	___

N/A - Not Applicable

WORKPLACE HEALTH PROTECTION PROGRAM ELEMENT CHECKLIST ON DELUGE SHOWER/EYEWASH FOUNTAIN

	Yes	No	Don't Know	N/A
. Provided with appropriate drain?	___	___	___	___
. Provided (where necessary) with a characteristic light that is visible at night?	___	___	___	___
. Standardized (arrangement, color, etc.) throughout the facility?	___	___	___	___
. Equipped with means to supply water at an acceptable temperature (60-95 F)?	___	___	___	___
. Installed with adequate height for:				
- Pull chain (70 inches)?	___	___	___	___
- Shower head (90 inches)?	___	___	___	___
- Eyewash fountain (40 inches)?	___	___	___	___
7. Is water flow rate adequate:				
. Shower (50 gallons per minute)?	___	___	___	___
. Eyewash (2.5 gallons per minute)?	___	___	___	___
8. Are equipment deficiencies promptly corrected?	___	___	___	___
9. Are dust caps kept on eyewash spigots?	___	___	___	___
10. Are records maintained of periodic operational checks?	___	___	___	___
11. Are bench mounted or free standing deluge showers/eyewash fountains available in the laboratory?	___	___	___	___
12. Is piping to outside deluge showers/eyewash fountains insulated to reduce heat load from the sun?	___	___	___	___

N/A - Not Applicable

WORKPLACE HEALTH PROTECTION PROGRAM ELEMENT CHECKLIST ON DELUGE SHOWER/EYEWASH FOUNTAIN

		Yes	No	Don't Know	N/A
13.	Are emergency deluge showers/ eyewash fountains in cold regions provided with means for supplying tempered water?	___	___	___	___
14.	Are portable emergency deluge shower/eyewash fountain facilities limited to temporary work locations or those where a piped water supply is impractical?	___	___	___	___
15.	Is the water that is available from portable emergency equipment fresh?	___	___	___	___
16.	Are eyewash bottles excluded from use except in areas where supplying water to emergency equipment presents a technical problem?	___	___	___	___
17.	Where bottled eyewash solution is provided, is an expiration date indicated on the container?	___	___	___	___

N/A - Not Applicable

WORKPLACE HEALTH PROTECTION PROGRAM ELEMENT
ON
HEARING PROTECTION

WORKPLACE HEALTH PROTECTION PROGRAM ELEMENT ON HEARING PROTECTION

OBJECTIVE

To provide information for the selection of appropriate hearing protection for use by personnel working in noisy areas.

BACKGROUND

Noise levels may exceed 85 dBA in some process units, off-site areas, maintenance shops, and other locations in petroleum or petrochemical facilities. Personnel may be exposed to these noise levels for sufficient time to result in hearing loss. Where feasible, engineering controls should be installed to reduce noise emissions from sources such as furnaces, motors, compressors, pressure reducing regulators, piping, and other sources. However, hearing damage risk situations may still exist. For example, personnel may work for varying periods of time in close proximity to noise sources, or use hand-held equipment which emits significant noise energy. For these tasks, the use of personal hearing protection is essential for reducing noise exposure since it is not always feasible to effectively control noise emissions from all sources.

The first step in determining the requirement for use of hearing protectors is to identify where hazardous noise exposure areas exist, whether personnel may receive an excessive noise exposure, and which sources could be controlled for reducing noise to which personnel are exposed. Where it is infeasible to reduce the noise emissions of significant noise sources through engineering means, or until such time as the controls are in place, the use of personal protection is necessary to reduce hearing damage risk.

A suggested action level for deciding whether hearing protective equipment is necessary is 85 dBA as an 8-hour time-weighted average (TWA) exposure (with 5 dBA doubling). In addition, if over 50 percent of the at-grade (general working area) noise levels in a process unit, off-site area, or other location equals or exceeds 85 dBA, it would be advisable to classify the location as a hearing protection zone. In addition, hearing protectors should be worn when working around, or with, noise producing equipment. Signs or labels should be posted on the equipment or in the immediate area of noise sources indicating this requirement.

Hearing protection available to personnel must have sufficient attenuation to reduce the noise level in the ear to below 85 dBA. All types of hearing protection are not equally effective for achieving this. The actual attenuation provided by a hearing protector varies for each noise source/frequency spectrum. The Noise Reduction Rating (NRR) for the type of protective device to be used should be evaluated relative to the noise characteristics of the source to which personnel are exposed to assure adequate attenuation is provided. The NRR value is typically supplied by the manufacturer for each type of hearing protector supplied.

The types of hearing protectors which are available include ear plugs, ear muffs, and canal caps. Each prevents some noise energy, which is transmitted through the air, from reaching the ear drum. More than one type of hearing protective device, which meets the attenuation needs for the noise for which it is to provide protection, should be made available to employees. Personnel should be allowed to choose the one which is most comfortable to them and they should be trained on the proper method of its use and maintenance.

Several methods are available for determining the adequacy of a specific hearing protective device. The simplest is referred to as the adjusted NRR method. It involves: 1) identification of the dBA value for the exposure/area/source for which the hearing protection is required, 2) obtaining the NRR value from the manufacturer for the type of protector to be used, 3) adjusting the NRR by subtracting 7 dBA from the NRR value, 4) subtracting this adjusted NRR value from the exposure level of the noise, and 5) comparing this value with 85 dBA, the target value. The protected exposure level should be less than 85 dBA. As an example of this procedure, a hearing protective device with an NRR of 25 dBA is to be used in an area where the noise level is 105 dBA. Would it provide adequate protection?

A-weighted TWA for work under a furnace:	105 dBA
"Adjusted" NRR value of device (25-7 = 18):	-18
Estimated protected exposure level:	87 dBA

This type of hearing protector would be inadequate for this task and have to be replaced by a more effective one to achieve the additional attenuation (e.g., 3 dBA) to get below 85 dBA. For some situations, use of insert type protectors and ear muffs simultaneously will suffice. Other methods for determining the effectiveness of hearing protective devices are discussed in the references.

Feedback from medical (or whomever does audiometric testing) is essential to determine if the personal protection is effective in preventing temporary or permanent hearing threshold shifts among users. If hearing threshold shifts are observed, an alternate type protector may be warranted and/or retraining of affected personnel on how to obtain effective protection with the device provided.

CONSIDERATIONS

The selection of a personal hearing protector for use by personnel who are excessively exposed to noise requires knowledge of the noise characteristics for which the device is to provide protection. Area, source, and personal noise exposure data are needed to make a selection of appropriate hearing protection. In the absence of such detailed information, measurement of the noise exposure levels (dBA values) and an estimate of the time and frequency of exposure can be used to evaluate the hazard potential and select appropriate hearing protection.

After the selection of the appropriate hearing protectors has been made, personnel should be given an opportunity to select the type most suitable to them, trained in the proper method of its use and maintenance (if appropriate), instructed as to where to use this personal protective equipment, and informed of the risks associated with failure to use it.

The effectiveness of the hearing protection for reducing noise exposure must be evaluated periodically. If a hearing shift is observed, or continues to develop in individuals required to use hearing protection, it is possible personnel are not using the equipment properly, or the type of device being used is not effective. These situations should be investigated and necessary steps taken to reduce hearing damage risk. This will likely require interaction between medical, industrial hygiene, supervision, the noise exposed individual, and others, as appropriate.

Hearing protection must be readily available to personnel. If ear muffs or canal caps are to be used for hearing protection it would be advisable to provide them to identified noise exposed personnel to carry on their person for use while working in high noise areas. Pre-molded insert type hearing protectors must be fitted and issued to the user. Foam type protectors should be available at the entry/exit of facility control rooms, at the shop, and other locations where hearing protection use is indicated.

REFERENCES

"Method For The Measurement Of Real-Ear Protection Of Hearing Protectors And Physical Attenuation of Earmuffs" (New York: American National Standards Institute, 1974).

Berger, E. H., "Attenuation Of Earplugs Worn In Combination With Earmuffs", E.A.R. Log-13 (Indianapolis, IN: Cabot Corp., March 1984).

Berger, E. H., "Methods of Measuring the Attenuation of Hearing Protection Devices", Journal of the Acoustic Society of America, 1986, pp. 1655-1687.

"List of Personal Hearing Protectors and Attenuation Data", Report No. 76-120 (Cincinnati, OH: United States Department of Health, Education, and Welfare, National Institute for Occupational Safety and Health, 1976),

Pekkarinen, J., "Industrial Impulse Noise, Crest Factor and the Effect of Earmuffs", American Industrial Hygiene Association Journal, October 1987, pp. 861-866.

Cluff, L. C., "Suggested Standard Hearing Protector Attenuation Function Relating to the NRR", American Industrial Hygiene Association Journal, December 1986, pp. 776-778.

WORKPLACE HEALTH PROTECTION PROGRAM ELEMENT CHECKLIST ON HEARING PROTECTION

		Yes	No	Don't Know	N/A
1.	Have measurements been made to determine the need for use of hearing protection in plant areas:				
	. Sound level measurements?	___	___	___	___
	. Noise contour maps?	___	___	___	___
	. Octave band analysis of noise sources/fields?	___	___	___	___
	. Personal noise dosimetry surveys?	___	___	___	___
2.	Have engineering controls been installed on significant noise sources that have been identified?	___	___	___	___
3.	Has hearing protection been selected on the basis of the characteristics of the noise to which individuals are exposed?	___	___	___	___
4.	Have appropriate signs been posted in areas/locations where hearing protective equipment should be worn?	___	___	___	___
5.	Have hearing protectors been made available to personnel whose 8-hour time-weighted average noise exposures are 85 dBA or more?	___	___	___	___
6.	Have personnel been trained as to the methods of proper use and maintenance of the hearing protection provided?	___	___	___	___
7.	Are reusable hearing protectors inspected periodically to ensure they are not defective?	___	___	___	___
8.	Is hearing protection (of the appropriate type) readily available to noise exposed personnel?	___	___	___	___

N/A - Not Applicable

WORKPLACE HEALTH PROTECTION PROGRAM ELEMENT CHECKLIST ON HEARING PROTECTION

		Yes	No	Don't Know	N/A
9.	Is hearing protection being worn by personnel working in areas where its use is indicated?	___	___	___	___
10.	Is there interaction between medical, industrial hygiene, supervision, and others regarding hearing threshold shifts and protective equipment use?	___	___	___	___
11.	Is consideration given to the use of two types of hearing protection simultaneously (where necessary) to reduce hearing damage risk?	___	___	___	___
12.	Has the variability of hearing protector attenuation been considered in selecting devices for use by personnel?	___	___	___	___
13.	Have published procedures been used for selection of personal hearing protection?	___	___	___	___
14.	Have high noise level sources to which employees may be exposed for an appreciable period of time (e.g., 15 minutes or more) been identified?	___	___	___	___
15.	Is the effectiveness of the hearing protection program evaluated periodically?	___	___	___	___
16.	Can personnel hear emergency signals while wearing hearing protection in noisy areas?	___	___	___	___
17.	Are employees encouraged to take hearing protection home if they engage in noisy activities there?	___	___	___	___

N/A - Not Applicable

WORKPLACE HEALTH PROTECTION PROGRAM ELEMENT
ON
OTHER PROTECTIVE EQUIPMENT

WORKPLACE HEALTH PROTECTION PROGRAM ELEMENT ON OTHER PROTECTIVE EQUIPMENT

OBJECTIVE

To provide information for the development, implementation, and maintenance of a protective equipment program which addresses workplace health concerns and aids in reducing exposure to workplace health hazards.

BACKGROUND

In the context of a workplace health protection program, personal protective equipment (PPE) includes any item or equipment which is worn or used to provide protection against exposure to physical, chemical or biological agents or ergonomic stressors. It includes gloves, aprons, face shield, boots, etc. Respiratory protective equipment is addressed as a separate program element.

Use of approved, properly fitted protective equipment is an acceptable approach for controlling exposure to workplace health hazards whenever engineering or administrative controls are infeasible or as an interim control measure until controls are implemented. Since many tasks carried out in the petroleum and petrochemical industries are of short duration, the use of personal protective equipment is often necessary to effectively control an exposure to a health hazard. Examples of such tasks include: the intermittent use of hand tools that emit high noise levels and for which personal hearing protection use is necessary to reduce hearing damage risk; use of eye, face and skin protection when hooking up lines to transfer/receive corrosive materials; wearing respiratory protection when there is the possibility that a contaminant may be released from a pump/line/vessel/valve, etc. on which one is working; handling chemicals that can be absorbed through the skin; use of protective clothing and respiratory protection when working with insulation; wearing thermal clothing while working in a cold environment; as well as others.

General information on the types of protective equipment that should be available to personnel involved in various tasks is incorporated in the MSDSs. These should be accessible to employees at their workplace. In addition, technical literature can be referred to for information on the appropriate equipment to be used with specific chemicals, for various tasks, or in different environments. When unsure of what protective equipment to use, guidance should be obtained through discussion with the site safety/health professional.

CONSIDERATIONS

The protective equipment to be made available to personnel for providing workplace health protection must be selected for the specific application or task so that the required protection is provided. Considerations in equipment selection include the form of the contaminant (e.g., vapor, gas, mist, dust, etc.) and its concentration; temperature of the environment; route of exposure (e.g., by inhalation, through the skin, or by ingestion); toxicity of the substance; protection afforded by the equipment; length of exposure; comfort factors; ruggedness of the equipment; fitness of the wearer; heat stress potential; equipment fit to enable the wearer to perform the task; etc.

When made available, the equipment must be a type that is approved by a competent authority, or meets an appropriate standard. Factors such as worker preference, acceptability, and comfort should be given consideration over cost, but not over protection effectiveness when selecting personal protective equipment.

Training must be provided personnel on the proper use, care and maintenance of the equipment. Measures must be taken to ensure that personnel are aware of the reasons for using the equipment, and supervisors must enforce its use when its need is indicated. Certain personal protective equipment, such as respiratory protection, should be assigned to the individual, whenever possible. Periodic checks should be carried out to ensure personnel are using the appropriate personal protective equipment properly.

Personal protective equipment should be effectively inventoried, stored and maintained to ensure an adequate supply is available. The effort to accomplish this during turnaround (maintenance) periods can be significant and additional support may be needed. Equipment should be cleaned and inspected after use, then properly stored to ensure it remains clean and does not deteriorate between usage periods (due to oxidation, photolysis, heat, creasing, etc.). Personnel who clean protective equipment that has been used should be advised if it has been contaminated with a hazardous material and, if necessary, provided appropriate personal protective equipment when handling and cleaning it. If a commercial laundry is employed to clean the equipment, that management should be informed of the exposure potential resulting from the contaminated equipment.

Emergency equipment should be located throughout the facility as needed. It should be periodically inspected and maintained so that it is available for immediate use. It should be replaced immediately after use. In no case should emergency equipment be used for non-emergency situations.

The protective equipment program should be reviewed periodically to ensure it is effective in meeting needs. Whenever possible, only equipment that has been approved by a testing authority should be made available to personnel. When appropriate, biological testing of personnel should be carried out to evaluate the effectiveness of the total exposure control program, including the personal protective equipment component.

REFERENCES

Berardinelli, S. P., et al, "A Portable Chemical Protective Clothing Test Method: Application at a Chemical Plant", American Industrial Hygiene Association Journal, September 1987, pp. 804-813.

Shulte, H. F., "Personal Protective Devices", The Industrial Environment - Its Evaluation and Control (Cincinnati, OH: National Institute for Occupational Safety and Health, 1973), pp. 519-531.

Perkins, J. L., "Chemical Protective Clothing: I. Selection and Use", Applied Industrial Hygiene, November 1987, pp. 222-230.

"Inspection for Accident Prevention in Refineries", API Publication 2002 (Washington, D. C.: American Petroleum Institute, 1984).

"Safe Maintenance Practices in Refineries", API Publication 2002 (Washington, D.C.: American Petroleum Institute, 1984).

"Personal Protective Equipment", Accident Prevention Manual for Industrial Operations, 7th Edition, Chapter 19 (Chicago, IL: National Safety Council, 1974), pp. 465-527.

Forsberg, K. and S. Z. Mansdorf, "Quick Selection Guide to Chemical Protective Clothing" (New York: Van Nostrand Reinhold, 1989).

Lundin, A. M., "Respiratory Protective Equipment", Fundamentals of Industrial Hygiene, 2nd Edition, Chapter 23 (Chicago, IL: National Safety Council, 1979), pp. 709-756.

Johnson, J. S., and K. J. Anderson, Ed., "Chemical Protective Clothing", Volumes 1 and 2 (Akron, OH: American Industrial Hygiene Association, 1990).

Johnson, J. S., and K. J. Anderson, Ed., "Chemical Protective Clothing", Vol. 2 (Akron, OH: American Industrial Hygiene Association, 1990).

WORKPLACE HEALTH PROTECTION PROGRAM ELEMENT CHECKLIST ON OTHER PROTECTIVE EQUIPMENT

		Yes	No	Don't Know	N/A
1.	Is personal protective equipment (PPE) selected for specific jobs/tasks on a case-by-case basis?	___	___	___	___
2.	Factors considered in PPE selection:				
	. Concentration of contaminant?	___	___	___	___
	. Routes of exposure to contaminant?	___	___	___	___
	. Toxicity of material?	___	___	___	___
	. Effectiveness of PPE (e.g., protection factor)?	___	___	___	___
	. Length of exposure?	___	___	___	___
	. Nature of the task/job?	___	___	___	___
	. Comfort factors of wearer?	___	___	___	___
	. Fitness of the wearer?	___	___	___	___
	. Equipment fit?	___	___	___	___
3.	Are employee considerations a part of PPE selection:				
	. Preference?	___	___	___	___
	. Acceptability?	___	___	___	___
	. Comfort?	___	___	___	___
4.	Are personnel given information as to:				
	. Why PPE is being provided?	___	___	___	___
	. How to properly use PPE?	___	___	___	___
	. How to clean PPE?	___	___	___	___
	. How to maintain/store PPE?	___	___	___	___
5.	Is PPE:				
	. Properly stored?	___	___	___	___
	. Inspected after usage?	___	___	___	___
	. Effectively cleaned?	___	___	___	___
	. Effectively maintained?	___	___	___	___
6.	Do supervisory personnel enforce PPE use where equipment need has been identified?	___	___	___	___

N/A - Not Applicable

WORKPLACE HEALTH PROTECTION PROGRAM ELEMENT CHECKLIST ON OTHER PROTECTIVE EQUIPMENT

	Yes	No	Don't Know	N/A
7. Are personnel who handle, clean, and maintain used PPE aware of the exposure potential that may exist:				
. Safety personnel?	___	___	___	___
. Storehouse personnel?	___	___	___	___
. Laundry (off-site)?	___	___	___	___
8. Are emergency PPE:				
. Effectively located throughout the site?	___	___	___	___
. Periodically inspected?	___	___	___	___
. Inspection records available?	___	___	___	___
. Replaced immediately following use?	___	___	___	___
. Not permitted for use in non-emergency situations?	___	___	___	___
9. Is the PPE program reviewed periodically to ensure its effectiveness?	___	___	___	___
10. Are jobs/tasks requiring use of PPE periodically reviewed to determine the adequacy of the PPE used?	___	___	___	___
11. If applicable, is biological testing performed periodically on PPE users to determine the effectiveness of equipment and the PPE program?	___	___	___	___
12. Is replacement PPE readily available to employees?	___	___	___	___
13. Are PPE needs for turnaround activities effectively addressed beforehand?	___	___	___	___
14. Is contact lens use disallowed where exposure to a corrosive/ irritant substance may occur?	___	___	___	___

N/A - Not Applicable

WORKPLACE HEALTH PROTECTION PROGRAM ELEMENT CHECKLIST ON OTHER PROTECTIVE EQUIPMENT

		Yes	No	Don't Know	N/A
15.	Is glove material selected based on the chemical(s) to which exposure may occur?	___	___	___	___
16.	Based on observations of personnel, is the proper personal protective equipment being used?	___	___	___	___
17.	Based on observations of personel, is personal equipment being used properly?	___	___	___	___
18.	Is special protective equipment available for spill clean up (as necessary)?	___	___	___	___
19.	Are deluge showers/eyewash fountains designed and operated in keeping with applicable specifications?	___	___	___	___
20.	Are signs indicating need for personal protective equipment use effective?	___	___	___	___

N/A - Not Applicable

WORKPLACE HEALTH PROTECTION PROGRAM ELEMENT
ON
RESPIRATORY PROTECTION

WORKPLACE HEALTH PROTECTION PROGRAM ELEMENT ON RESPIRATORY PROTECTION

OBJECTIVE

To provide information to ensure that respiratory protection programs are developed, implemented, and maintained to meet the needs of facility personnel for carrying out work assignments without excessive risk when respiratory protective equipment use is warranted.

BACKGROUND

The application of engineering measures is the preferred method for controlling employee exposure to airborne contaminants. This is typically achieved in petroleum and petrochemical operations through the use of pump/compressor seals and gaskets, closed systems, exhaust ventilation, directing releases away from work areas, etc. There are occasions, however, when the use of respiratory protection is necessary to protect the health of personnel. Thus, it is essential that an effective respiratory protection program be established and maintained.

CONSIDERATIONS

Where there is a need for use of respiratory protection, the employer is responsible for establishing and maintaining a respiratory protective equipment program that is effectively administered and meets employee needs. The details of the program that is developed and implemented should be written and a program administrator identified.

The selection of respirators must be made on the basis of the hazards to which personnel are exposed. Only approved equipment should be available to personnel. Respirator users should be trained in the proper use of the protective equipment, be fit tested, and made aware of equipment limitations. Personnel who may potentially use respiratory protection should be clean shaven and be provided a medical examination to ensure their fitness to use the equipment. The training, fit testing, and medical examination should be repeated periodically.

Procedures must be developed to clean and inspect respirators after each use and provide maintenance to ensure equipment reliability. A monthly inspection program should be implemented for emergency use respirators. In addition, these should be inspected after each use. Users of respirators which, according to the manufacturers' instructions, can be used for a specified time period, should indicate the time periods of use on the canister and seal the canister after each use.

An evaluation of the respirator program should be conducted periodically to ensure its continued effectiveness.

Air supplied to workers via self-contained or air supplied systems must meet the specifications for acceptable breathing air. The couplings on airlines for providing breathing air to personnel should be incompatible with fittings of other airlines (process/instrument), or those of non-respirable gas distribution systems (nitrogen, hydrogen, etc.).

Respiratory protection procedures/practices should be established for confined space entry and for immediately dangerous to life or health (IDLH) situations. Additional respirator and hazard awareness training, as well as emphasis of accepted procedures/work practices, should be provided personnel involved in confined space work or where sour materials (e.g., some crude units, desulfurization units, sulfur recovery unit gas streams, etc.) are handled.

REFERENCES

"Practices for Respiratory Protection", ANSI Z88.2 (New York: American National Standards Institute, 1980).

"Respiratory Use - Physical Qualifications for Personnel", ANSI Z88.6 (New York: American National Standards Institute, 1984).

"Lundin, A. M., "Respiratory Protective Equipment", Fundamentals of Industrial Hygiene, 2nd Edition, Chapter 23 (Chicago, IL: National Safety Council, 1979), pp. 709-756.

"Guide to Industrial Respiratory Protection", (Cincinnati, OH: National Institute for Occupational Safety and Health, 1987).

Ryhans, G. S. and D. S. L. Blackwell, "Practical Guide to Respirator Usage in Industry" (Boston, MA: Butterworth Publishers, 1985).

Johnson, J. S. and K. J. Anderson, Ed. "Chemical Protective Clothing", Vol.1 (Akron, OH: American Industrial Hygiene Association, 1990).

WORKPLACE HEALTH PROTECTION PROGRAM ELEMENT CHECKLIST ON RESPIRATORY PROTECTION

		Yes	No	Don't Know	N/A
1.	Is there a written respirator program?	___	___	___	___
2.	Have respirators been selected based on potential exposure hazards?	___	___	___	___
3.	Are respirators that are provided to personnel approved for their use/application?	___	___	___	___
4.	Are employees trained in the use of the respirators they may be required to use and instructed as to their limitations?	___	___	___	___
5.	Are records available on personnel who have been trained in respirator use?	___	___	___	___
6.	Have employees who may be required to use respirators been medically certified?	___	___	___	___
7.	Are employees medically re-certified periodically for respirator use?	___	___	___	___
8.	Are lists of employees in the respiratory medical surveillance program kept up to date with additions/deletions made promptly?	___	___	___	___
9.	Are respirators cleaned, inspected, and maintained on a regular basis?	___	___	___	___
10.	Are respirators stored such that they are clean, sanitary, and ready for use?	___	___	___	___
11.	Are emergency use respirators inspected on a monthly basis and after each use?	___	___	___	___
12.	Is there a written procedure for inspection of respirators?	___	___	___	___

N/A - Not Applicable

WORKPLACE HEALTH PROTECTION PROGRAM ELEMENT CHECKLIST ON RESPIRATORY PROTECTION

		Yes	No	Don't Know	N/A
13.	Are personnel who issue respirators provided appropriate training for selecting the proper type for the application?	___	___	___	___
14.	Are records kept of emergency use respirator inspections?	___	___	___	___
15.	Have personnel who inspect respirators been properly trained?	___	___	___	___
16.	Is there a clean face policy in effect?	___	___	___	___
17.	Are engineering controls adequate throughout the facility so there is no need for routine use of respirators?	___	___	___	___
18.	Is the respirator program reviewed annually to ensure its effectiveness?	___	___	___	___
19.	Are the results of periodic respirator program reviews:				
	. Documented?	___	___	___	___
	. Communicated to management?	___	___	___	___
20.	Are action plans promptly developed to correct respirator program deficiencies?	___	___	___	___
21.	Are individuals who have difficulty using a respirator due to a possible underlying medical condition promptly referred to medical for evaluation?	___	___	___	___
22.	Are personnel periodically retrained on respirator use?	___	___	___	___
23.	Are breathing air compressors used?	___	___	___	___
24.	Is breathing air from compressors:				
	. Supplied directly to users?	___	___	___	___
	. Cleaned to remove contaminants?	___	___	___	___

N/A - Not Applicable

WORKPLACE HEALTH PROTECTION PROGRAM ELEMENT CHECKLIST
ON
RESPIRATORY PROTECTION

	Yes	No	Don't Know	N/A
. Periodically tested to ensure it meets breathing air specifications?	___	___	___	___
25. Is the breathing air compressor provided with a high temperature and CO alarm in the system?	___	___	___	___
26. Is all breathing air (purchased and that provided by in-house equipment) tested to ensure it meets appropriate specifications?	___	___	___	___
27. Are emergency respirators not permitted to be used in routine use applications?	___	___	___	___
28. Are the locations of emergency use respirators adequately identified?	___	___	___	___
29. Are breathing air couplings incompatible with those on other air/gas distribution systems?	___	___	___	___
30. Are breathing air cylinders properly identified?	___	___	___	___
31. Is a back-up individual present and properly equipped when a person is working in a potential IDLH situation?	___	___	___	___
32. Are personnel fit tested if they are required to use respiratory protective equipment? Check which: Qualitative Quantitative	___	___	___	___
33. Are there adequate provisions for personnel who must wear glasses while wearing respiratory protection?	___	___	___	___
34. Is contact lens use disallowed by personnel while wearing full facepiece type respiratory protection?	___	___	___	___

N/A - Not Applicable

WORKPLACE HEALTH PROTECTION PROGRAM ELEMENT CHECKLIST ON RESPIRATORY PROTECTION

		Yes	No	Don't Know	N/A
35.	Is contact lens use disallowed by personnel wearing half-mask respiratory proction where dusts, mists, fumes, or corrosive materials may be present?	___	___	___	___
36.	Are respirators that may be used by more than one person effectively cleaned and disinfected after each use?	___	___	___	___
37.	Are respirator cartridges/canisters sealed during storage?	___	___	___	___
38.	Are repairs of respiratory protective equipment made by qualified personnel?	___	___	___	___
39.	Is respiratory protection being used properly by facility personnel?	___	___	___	___

N/A - Not Applicable

6. SPECIAL MATERIALS

WORKPLACE HEALTH PROTECTION PROGRAM ELEMENT
ON
ALKYL LEAD COMPOUNDS

WORKPLACE HEALTH PROTECTION PROGRAM ELEMENT ON ALKYL LEAD COMPOUNDS

OBJECTIVE

To provide information on the handling procedures and work practices that are appropriate to minimize exposure to lead compounds during lead blending, tank cleaning, sludge disposal, and tank dismantling operations.

BACKGROUND

Alkyl lead compounds, such as tetraethyl (TEL) and tetramethyl (TML) lead, have been used for years in refining operations as a motor gasoline and aviation fuel additive. TEL and TML are extremely toxic. Adverse health effects can occur as a result of the inhalation or skin absorption of these materials. Handling procedures/work practices, that have been identified by DuPont, Ethyl, and Octel, the primary suppliers of these additives, must be adhered to in order to minimize worker exposure risk and contamination of the environment. The practices to be followed include the use of protective clothing; adherence with recommended handling procedures and special work practices during blending, tank cleaning, and disposal of sludge; and the proper disposition of scale/steel from leaded gasoline storage tanks.

The use of alkyl lead additives is being phased out as a motor gasoline additive. They are, however, still in use in some locations as a motor fuel additive and routinely as an additive in aviation gasoline. The health concerns associated with sludge handling from blending equipment and gasoline storage tanks, as well as for the disposition of the steel from these facilities/operations, will continue to be a concern for a number of years.

CONSIDERATIONS

The procedures/work practices that have been established for the use of alkyl lead compounds were developed, for the most part, by the manufacturers of these additives. Additional information has been developed by lead alkyl users and petroleum industry trade associations. They are detailed and specific with respect to workplace and environmental health concerns and are based on worldwide experience in the use of the alkyl lead compounds. It is essential that the manufacturers' and trade associations' literature be reviewed and the alkyl lead handling procedures/work practices be developed and adhered to based on that information. Some of the key requirements that need to be implemented include the following:

General

Blending personnel, knock-test lab employees, tank cleaners, and facility maintenance personnel may be exposed to alkyl lead compounds in the course of their work. They must be aware of the potential adverse health effects and of the procedures/work practices to be followed to minimize their exposure to the alkyl lead compounds.

Personnel should be aware that alkyl lead compounds used in gasoline blending present an exposure potential by inhalation and skin absorption. Appropriate protective equipment, including coveralls, impervious gloves, boots, and respiratory protection may be necessary for use by personnel exposed to these compounds.

Personnel potentially exposed to alkyl lead compounds should be trained in the use of appropriate control procedures and the measures to take if skin contact with an alkyl lead compound occurs.

Alkyl lead contaminated equipment should be effectively flushed with kerosene and water before removing it from service for maintenance. In some locations, a solution of potassium permanganate is used to wash equipment before working on it.

A urinary lead monitoring program should be established for personnel exposed to alkyl lead compounds in their work.

Laboratory

Alkyl lead compounds are to be stored and handled inside a mechanically exhaust ventilated hood. The exhaust hood must be properly designed and the face velocity should meet that recommended by the manufacturer. Personnel should be provided the appropriate protective equipment and required to wear it (gloves and face shield while handling these materials). TEL/TML must be properly stored and there should be no evidence of spillage in the hood. The work area should be kept clean. If spills occur they should be promptly cleaned up.

Laboratory personnel who work with TEL/TML should be included in the organic lead biological monitoring program if there is risk of skin contact or exposure by inhalation.

Gasoline Blending

Personnel working in lead blending operations must be

acquainted with, and follow established operating procedures. These should be posted in the blending area.

The blending facility should be kept locked when not in use. There should be no evidence of TEL/TML spills and the appropriate protective equipment should be available and worn by personnel working there. Drums (if present) should be properly stored. Air purifying respirator use period records (date/hours) should be maintained and the cartridges/canisters kept sealed per the vendors instructions.

Kerosene should be available at the change/locker room for removal of alkyl lead material from the skin if contamination occurs. In addition, personnel should take a shower following the blending operation.

Tank Cleaning

Tanks containing leaded product, and those which previously contained leaded product, should be identified with a sign indicating this fact.

Training on the appropriate procedures and work practices should be provided personnel involved in leaded gasoline storage tank cleaning. Personnel should be medically approved, and provided hazard awareness training appropriate for this work. The proper protective equipment should be available and used (rubber boots and gloves, coveralls, self-contained breathing apparatus or supplied-air respirator with escape pack) by tank cleaning personnel, when appropriate. Access to the work area should be limited to tank cleaning personnel.

A test must be conducted to determine the airborne vapor concentration of the alkyl lead compound(s) prior to vessel entry by personnel. Alkyl lead manufacturers have published guidelines on permissible exposure times that are appropriate for various airborne concentrations of alkyl lead vapor. If it is necessary to be in a tank for a period longer than published for a given concentration, then appropriate respiratory protection should be worn. Compressors for supplied air systems, if used, are to be located such that emissions from the storage tank and compressor engine do not enter the air inlet. Equipment used in tank cleaning should be effectively cleaned/decontaminated when work is completed.

Personnel involved in tank cleaning should shower at the end of the work and be included in the alkyl lead biological monitoring program.

Tank Inspection/Repair

Leaded tank entry procedures should be followed whenever it is necessary to enter a leaded gasoline storage tank that has not been determined to be lead free. Tanks that require repair of bottom leaks or work on other inside sections of a leaded tank (such as the roof or pontoons) should be cleaned to bare metal before the repair is made. Appropriate respiratory protection and other protective equipment should be worn during these operations to reduce exposure to lead compounds.

Sludge Disposal

Weathering, for a number of years, was a widely employed method for disposal of leaded gasoline sludge. It may yet be in use in some locations since it was considered the preferable method by some alkyl lead manufacturers. The procedure involved spreading the sludge out (about 3 inches deep) on a concrete pad and allowing the sun and weather to evaporate some of the material and decompose the remaining TEL/TML to inorganic lead. The weathered (dry) sludge was transferred into drums for disposal as hazardous waste at an approved disposal site in accordance with applicable regulations. Respirators and other protective equipment were typically worn when handling weathered sludge. Burial of sludge within the tank diked area, or other marked location has not been employed as a disposal method for many years.

Methods of sludge disposal that are presently employed are almost universally controlled by regulations. Adherence with these requirements is essential.

Tank Dismantling

There are restrictions for the use of steel from leaded gasoline storage tanks and other equipment that has been in contact with alkyl lead compounds or leaded fuels. These use limitations have been defined by alkyl lead manufacturers and petroleum trade associations. The concern is to prevent the use of this steel in applications where it may come in contact with foodstuff for use by animal or man.

Steel from equipment in alkyl lead service that has scale present on its surface must be cleaned to bare metal if it is to be reused. Such cleaning is always necessary for both the bottom tank ring and the tank bottom of leaded gasoline storage tanks. In addition, other parts of tanks that were in leaded gasoline service that show significant scaling should be cleaned to bare metal if the steel is to be reused. The alternative is to bury the tank steel if it is not to be cleaned as recommended.

The purchaser of the scrap steel from the tank must be informed of the lead contamination potential and of the restrictions for its use.

REFERENCES

Sherwood, R. J., "Safe Demolition and Disposal of a Redundant Alkyl Lead Storage Tank", American Industrial Hygiene Association Journal, August 1990, pp. 405-415.

"Cleaning Petroleum Storage Tanks", API Publication 2015, Fourth Edition (Washington, D. C.: American Petroleum Institute, 1991).

"Guide for Controlling the Lead Hazard Associated With Tank Entry and Cleaning", API Publication 2015A (Washington, D. C.: American Petroleum Institute, 1982).

"Preparing Tank Bottoms for Hot Work", API Publication 2207 (Washington, D. C.: American Petroleum Institute, 1982).

"Dismantling and Disposing of Steel From Tanks Which have Contained Leaded Gasoline", API Publication 2202, Third Edition (Washington, D. C.: American Petroleum Institute, 1982).

"Lead Alkyl Antiknock Compounds; Laboratory Practice", Octel Instructional Publication No. 6, 1988.

Ethyl Antiknock Service Manual, Ethyl Corporation, 1980.
"A Guide For Medical Examiners And Consultants Responsible For The Medical Supervision Of Handlers Of Tetraethyl And Tetramethyl Lead At Oil Refineries", Medical Department, Ethyl Corporation, October 1984.

WORKPLACE HEALTH PROTECTION PROGRAM ELEMENT CHECKLIST ON ALKYL LEAD COMPOUNDS

		Yes	No	Don't Know	N/A
1.	Is TEL/TML received by:				
	. Drum?	___	___	___	___
	. Tank wagon?	___	___	___	___
	. Rail car?	___	___	___	___
	. Ship?	___	___	___	___
2.	Is the number of personnel involved in working where exposure to alkyl lead compounds may occur kept to a minimum?	___	___	___	___
3.	Are only authorized personnel permitted in alkyl lead storage and blending areas?	___	___	___	___
4.	Is access to the alkyl lead storage area limited when receiving a shipment of alkyl lead compound?	___	___	___	___
5.	Are lead-in-air test kits available and used to test for the presence of alkyl lead compounds, when appropriate?	___	___	___	___
6.	Are lead-in-air test kits within expiration date?	___	___	___	___
7.	Are the following measures in place at the alkyl lead storage area:				
	. Restricted access?	___	___	___	___
	. No evidence of spills?	___	___	___	___
	. Means available for prompt spill clean-up?	___	___	___	___
8.	Is there a regular shower, a supply of kerosene, and an eyewash fountain readily available to personnel at the alkyl lead storage/blending area?	___	___	___	___
9.	Are alkyl lead drums stored according to manufacturer's instructions?	___	___	___	___

N/A - Not Applicable

WORKPLACE HEALTH PROTECTION PROGRAM ELEMENT CHECKLIST
ON
ALKYL LEAD COMPOUNDS

	Yes	No	Don't Know	N/A
10. Is protective equipment use adequate when there is a potential for exposure to alkyl lead compounds:				
. Boots?	___	___	___	___
. Gloves?	___	___	___	___
. Coveralls?	___	___	___	___
. Respiratory protection?	___	___	___	___
11. Are personnel who are potentially exposed to alkyl lead compounds trained regarding:				
. Hazard awareness?	___	___	___	___
. Lead handling procedures/ work practices?	___	___	___	___
. Respiratory protection use?	___	___	___	___
. Confined space entry (when appropriate)?	___	___	___	___
. Emergency procedures?	___	___	___	___
12. Do alkyl lead exposed personnel who have a potential to contact these materials shower after work is completed?	___	___	___	___
13. Is the lead blending facility inspected periodically to assure:				
. Good housekeeping is maintained?	___	___	___	___
. Respiratory protective equipment is in good condition?	___	___	___	___
. No evidence of alkyl lead leakage?	___	___	___	___
. Showers are working properly?	___	___	___	___
. Kerosene supply is available?	___	___	___	___
Date of last inspection? ________________				
14. Is the alkyl lead laboratory hood ventilation periodically evaluated?	___	___	___	___
15. Is the laboratory TEL/TML hood ventilation adequate?	___	___	___	___
16. Is the TEL/TML hood housekeeping acceptable?	___	___	___	___

N/A - Not Applicable

WORKPLACE HEALTH PROTECTION PROGRAM ELEMENT CHECKLIST ON ALKYL LEAD COMPOUNDS

		Yes	No	Don't Know	N/A
17.	Is TEL/TML properly stored in the laboratory?	___	___	___	___
18.	Do laboratory personnel who handle TEL/TML use the appropriate protective equipment?	___	___	___	___
19.	Is equipment in alkyl lead service (valves, piping, gaskets, etc.) that is to be discarded:				
	. Put in a specifically designated scrap pile?	___	___	___	___
	. Returned to lead supplier?	___	___	___	___
	. Buried?	___	___	___	___
	. Heated to red heat or cleaned to bare metal before burial?	___	___	___	___
	. Heated to red heat before classified as scrap destined for smelting and restricted use?	___	___	___	___
20.	Are copies available of the references cited in this document?	___	___	___	___
21.	Is appropriate equipment available for providing effective ventilation of leaded fuel storage tanks before and during cleaning?	___	___	___	___
22.	Is monitoring equipment available to determine if a tank is lead free?	___	___	___	___
23.	Is a standby person present when entry is made into a leaded gasoline tank for testing, inspection, work, etc.?	___	___	___	___
24.	Is appropriate respiratory protection and other protective equipment available for leaded gasoline tank entry when the need is indicated?	___	___	___	___

N/A - Not Applicable

WORKPLACE HEALTH PROTECTION PROGRAM ELEMENT CHECKLIST ON ALKYL LEAD COMPOUNDS

		Yes	No	Don't Know	N/A
25.	Is protective equipment (hoses, facepieces, boots, etc.) effectively cleaned on completion of alkyl lead work or at the end of a shift?	___	___	___	___
26.	Is there a shower facility available to personnel working in tanks that may contain alkyl lead compounds?	___	___	___	___
27.	Is leaded gasoline sludge disposed by:				
	. Burying?	___	___	___	___
	. Burning?	___	___	___	___
	. Incinerating?	___	___	___	___
	. Weathering?	___	___	___	___
	. Other procedure (__________)?	___	___	___	___
28.	Is the leaded gasoline sludge weathering site located in a remote area with restricted access?	___	___	___	___
29.	Are personnel who spread leaded sludge for weathering provided appropriate protective equipment to wear?	___	___	___	___
30.	Is the equipment used to spread/ work the sludge at the weathering site washed after each period of use?	___	___	___	___
31.	Is weathered sludge disposed in an acceptable manner?	___	___	___	___
32.	Are employees who use or handle alkyl lead materials included in the organic lead medical surveillance program?	___	___	___	___

N/A - Not Applicable

WORKPLACE HEALTH PROTECTION PROGRAM ELEMENT CHECKLIST ON ALKYL LEAD COMPOUNDS

	Yes	No	Don't Know	N/A
33. Are individuals who may be exposed to alkyl lead compounds included in the respirator medical surveillance program?	___	___	___	___
34. Is appropriate respiratory protective equipment worn by personnel when:				
. Removing scale from inside a leaded gasoline storage tank?	___	___	___	___
. Drumming lead containing scale?	___	___	___	___
. Welding/burning inside a leaded gasoline storage tank?	___	___	___	___
. Abrasive cleaning inside a leaded gasoline storage tank?	___	___	___	___
. Turning over sludge at the weathering area?	___	___	___	___
. Drumming weathered sludge for disposal?	___	___	___	___
35. Are personnel who may be exposed to inorganic lead associated with the weathering operation included in a medical surveillance program?	___	___	___	___

N/A - Not Applicable

WORKPLACE HEALTH PROTECTION PROGRAM ELEMENT ON AROMATIC PROCESS OILS

WORKPLACE HEALTH PROTECTION PROGRAM ELEMENT ON AROMATIC PROCESS OILS

OBJECTIVE

To provide information for reducing employee exposure to aromatic process oils and identify measures for reducing the likelihood of adverse health effects to occur where there is a potential for exposure to these compounds.

BACKGROUND

Aromatic process oil is a term used to designate the high-boiling streams that are produced as the result of catalytic or thermal cracking, or by the extraction of raw lube distillates. The aromatic process oils are complex mixtures of paraffinic, naphthenic and aromatic hydrocarbons in the boiling range of 500-1000 F (260-540 C). The molecular structures that are of most concern in these oils are the polycyclic aromatic compounds made up of four or more rings.

Aromatic process oils have been demonstrated to have carcinogenic potential in animal testing. For example, it has been determined that the bottom stream from the catalytic cracking process produces skin cancer in animals as a result of repeated and prolonged contact with this liquid. Toxic effects have been observed when other high boiling materials, which contain multi-ringed aromatic compounds, have been tested on animals. It has been demonstrated that the multi-ringed compounds of health concern concentrate in those petroleum fractions which boil above approximately 700 F. As examples, heavy gas oils, aromatic extracts from lube oil processing, heavy and light coker gas oil, steam cracked tar, heavy vacuum gas oil, as well as some other products and intermediates produce skin cancer in animal tests.

CONSIDERATIONS

Streams which contain aromatic process oils should be identified so that appropriate measures can be taken to minimize contact and the potential for adverse health effects to occur. Personnel who may be exposed to these materials should be aware of their presence, their potential adverse health effect, where exposure may occur, measures to take to prevent contact, and the method of removing these materials from the skin if contact does occur. If skin effects are observed, the individual should promptly report this to medical.

Personnel potentially exposed to high boiling aromatic oils should be provided appropriate protective clothing for minimizing contact potential. This may include impervious gloves, boots, and coveralls. A face shield is indicated for some tasks, such as sampling a hot process stream that contains an aromatic process oil.

Leaking equipment should be repaired promptly to prevent or minimize surface contamination, as well as the release of mists and vapors. If a release occurs, surface contamination should be contained and promptly cleaned up. Some vapors can condense and, depending upon the amount released, result in a haze in the area and a potential for excessive exposure of personnel.

Personnel potentially exposed to aromatic process oils should be instructed to:

- Perform work in well ventilated areas, where possible
- Use oil impervious protective clothing
- Avoid all personal contact with these materials
- Promptly dry wipe the skin if contact occurs, cleanse the affected area with waterless hand cleaner, and follow by washing thoroughly with soap and water
- Remove contaminated clothing and launder it before reuse

Maintenance supervisors should require their personnel to adhere to accepted work practices, use recommended protective clothing, and practice good personal hygiene. When indicated, additional protective clothing and respiratory protection should be provided. Equipment to be worked on should be drained and flushed before maintenance is carried out on it.

Personnel potentially exposed to aromatic process oils should shower at the end of their work, particularly if the tasks carried out were considered dirty.

Process units, furnaces, tanks, pumps and other equipment which contain, or process aromatic process oils should be identified with a sign, and personnel alerted to the presence of these contaminants. Wastes containing aromatic process oils must be properly disposed according to the material's classification.

REFERENCES

"IARC Monographs on the Evaluation of Carcinogenic Risks to Humans - Certain Polycylic Aromatic Hydrocarbons and Heterocyclic Compounds", Volume 3 (Lyon, France: International Agency for Research on Cancer, 1973).

Lesage, J., et al, "Evaluation of Worker Exposure to Polycyclic Aromatic Hydrocarbons," American Industrial Hygiene Association Journal, September 1987, pp. 753-759.

WORKPLACE HEALTH PROTECTION PROGRAM ELEMENT CHECKLIST ON AROMATIC PROCESS OILS

		Yes	No	Don't Know	N/A
1.	Have streams/units which contain aromatic process oils been identified in operations?	___	___	___	___
2.	Are personnel who may be exposed to aromatic process oils aware of the hazard potential of these materials?	___	___	___	___
3.	Have locations, operations, and equipment which contain aromatic process oils been effectively identified and this information communicated to personnel?	___	___	___	___
4.	Are appropriate hazard warning signs posted where exposure to aromatic process oils may occur?	___	___	___	___
5.	Have personnel who may be exposed to aromatic process oils been instructed on:				
	. Work practices to follow?	___	___	___	___
	. Protective clothing to wear?	___	___	___	___
	. Respiratory protective equipment to wear?	___	___	___	___
	. Methods to remove contamination from the skin?	___	___	___	___
	. The need for showering at the end of the shift?	___	___	___	___
	. The need for reporting skin effects promptly?	___	___	___	___
6.	Are aromatic process oil releases/ leaks promptly corrected/repaired?	___	___	___	___
7.	Is housekeeping acceptable in areas where aromatic process oils are handled/processed?	___	___	___	___
8.	Are there facilities available to practice good personal hygiene in areas/operations where exposure to aromatic process oils may occur?	___	___	___	___

N/A - Not Applicable

WORKPLACE HEALTH PROTECTION PROGRAM ELEMENT CHECKLIST ON AROMATIC PROCESS OILS

		Yes	No	Don't Know	N/A
9.	Is an effective skin cleaning agent available to personnel working in areas where contact with aromatic process oils may occur?	___	___	___	___
10.	Is appropriate respiratory protective equipment used in locations where aromatic process oils are released?	___	___	___	___
11.	Is extra effort made to remove aromatic process oil contamination from equipment before maintenance is done?	___	___	___	___
12.	Do permits issued for work in areas where aromatic process oils are present specify the appropriate personal protection to use?	___	___	___	___
13.	Do personnel shower at the end of work during which there was potential exposure to aromatic process oils?	___	___	___	___
14.	Are aromatic process oil containing wastes properly handled and disposed by personnel who are aware of their hazards?	___	___	___	___
15.	Does the periodic medical examination program for employees include a skin examination for personnel potentially exposed to aromatic process oils?	___	___	___	___

N/A - Not Applicable

WORKPLACE HEALTH PROTECTION PROGRAM ELEMENT
ON
ASBESTOS

WORKPLACE HEALTH PROTECTION PROGRAM ELEMENT ON ASBESTOS

OBJECTIVE

To provide information for ensuring that an effective asbestos exposure control program is in place which incorporates procedures/work practices for reducing or eliminating the potential for exposure of personnel to asbestos. [NOTE! Recently published information indicates that other fibrous materials (i.e., asbestos substitutes) may have toxic properties similar to asbestos. Materials such as mineral wool, refractory ceramic fibers, and other man-made mineral fibers should be handled such that exposure to these is minimized.]

BACKGROUND

Asbestos-containing materials (ACM) have been used in industrial operations as insulation, gaskets, fire protection, in construction (asbestos concrete board), as pump seals, and in other applications. Some have been, and still are, a source of fiber release to the air. This is particularly true during application, maintenance, and removal of the ACM. Because the inhalation of asbestos fibers can produce serious adverse health effects, measures must be taken to reduce exposure to this contaminant. These may include engineering and work practice controls, as well as the use of personal protective equipment.

CONSIDERATIONS

In order to minimize or eliminate exposure of personnel to asbestos, the following measures need to be implemented and maintained:

- Identify, as well as is practical, where asbestos is present in the facility. This requires bulk sampling and analysis of samples by polarized light microscopy or other method. In the absence of definitive information, a suspect material should be handled as if it contains asbestos.

- Develop written procedures for controlling exposure to asbestos. These should apply to all work where there is a potential for personnel (employees, as well as contractors) to be exposed to asbestos fibers.

- Eliminate the use of asbestos-containing materials, where feasible.

- Establish an operations and maintenance (O&M) program which includes procedures/work practices that ensure in-place ACM will not be disturbed and release fibers during maintenance, remodeling, construction or abatement activities.

- Develop and implement work specifications/guidelines for handling asbestos-containing materials. These should include, but not necessarily be limited to the following:

  - Ensure compliance with applicable regulations and company guidelines.

  - Ensure that personnel performing ACM work are properly trained in the facility's asbestos handling procedures and are aware of all applicable regulations.

  - Ensure asbestos exposed personnel are provided appropriate physical examinations and are medically approved for such work.

  - Establish barriers at the work location to restrict access to authorized personnel only. Eating, drinking or smoking are not to be permitted in the work area.

  - Require personnel doing ACM work to wear appropriate protective equipment (e.g., disposable coveralls, hair and shoe coverings, respirator, and gloves).

  - Select approved type respiratory protection based on the expected or measured airborne fiber concentration. Ensure that personnel who may be required to wear respiratory protection are medically fit (i.e., approved for such work) and, as a minimum, are qualitatively fit tested.

  - Use wet procedures, where practical, when doing ACM removal work. The water should incorporate a wetting agent.

  - Maintain good housekeeping throughout the work area to contain ACM and prevent its spread to surroundings.

- Use enclosures to aid in preventing the spread of asbestos-containing material outside the work area. This can be accomplished by using plastic sheeting to enclose the work area and exhausting air from this enclosure to create a negative pressure so that air flows into it. The exhaust unit should be equipped with a high efficiency particulate air (HEPA) filter.

- Where practical, use a glove bag (properly labeled and sealed) for removal of insulation from piping. Use of a glove bag often negates the need for enclosing the work area.

- Carry out personal monitoring to evaluate employee exposure to asbestos fibers using an acceptable procedure, such as the NIOSH 7400 Method. Results can be used to determine if the respiratory protective equipment provided personnel is appropriate.

- Maintain records of exposure results.

- Obtain samples outside the containment area while asbestos abatement is proceeding in order to determine if the background fiber level has been exceeded and a breach of the containment has occurred.

- Ensure that samples are analyzed in a qualified laboratory.

- Ensure that surfaces are clear of asbestos-containing material before applying a substitute. For most situations, a sealant should be applied to fix remaining fibers before the replacement is put in place.

- Place waste asbestos-containing material, contaminated protective clothing, and scrap in properly identified bags for subsequent disposal. Double bagging is urged.

- Ensure waste is disposed in an approved site. Disposal records should be kept (e.g., volumes, number of bags, disposal location, dates, etc.).

- Ensure that the asbestos work area is clean before its return to unrestricted access and routine use. For enclosed work areas, this can be accomplished by an aggressive sampling program.

. Establish a hazard communication program to inform personnel potentially exposed to ACM of the exposure risk, as well as the procedures/work practices to which they should adhere when working with such material. As a part of this effort, information should be developed that can be provided to contractors who work on-site with asbestos-containing materials.

- Require that contractors bidding on work in which ACM will be disturbed or handled to submit job proposals for review well in advance of the date for selection of the successful bidder. Such proposals should be reviewed in detail by industrial hygiene, safety, environmental, operations, maintenance, etc. Deficiencies identified will require modification of the proposal before the contract is awarded. (A list of contractors who have demonstrated acceptable procedures/work practices, and have appropriate licenses should be maintained).

- Identify an individual or group to provide an oversight function during work with asbestos-containing material to ensure adherence with company guidelines.

REFERENCES

Wright, G. W., "The Pulmonary Effects of Inorganic Dust", Chapter 7, Patty's Industrial Hygiene and Toxicology - General Principles, Vol. 1, 3rd Rev. (New York: John Wiley & Sons, 1978), pp. 194-198.

Price, C. M., Ed., "Asbestos in Buildings, Facilities & Industry" (Rockville, MD: Government Institutes, Inc., 1987).

"Guidance for Controlling Asbestos-Containing Materials in Buildings" (Washington, D. C.: U. S. Environmental Protection Agency, 1985).

"Managing Asbestos in Place" (Washington, D. C.: U. S. Environmental Protection Agency, 1990).

NIOSH Manual of Analytical Methods, "Method 7400 - Fibers", Revision 2 (Cincinnati, OH: National Institute for Occupational Safety and Health, 1987).

"Airborne Asbestos Fiber Concentrations at Workplaces by Light Microscopy", Method #1, Membrane Filter Method (London: Asbestos International Association, 1979).

Grohlich, D. and W. C. Monson, "Get a New Angle on Asbestos Management", Safety and Health, February 1991, pp. 43-49.

Buford, J. L., "Selecting Your Asbestos Abatement Contractor", Professional Safety, February 1991, pp. 15-17.

Dement, J. M. and K. M. Wallingford, "Comparison of Phase Contrast and Electron Microscopic Methods for Evaluation of Occupational Asbestos Exposures", Applied Occupational and Environmental Hygiene, April 1990, pp. 242-247.

"On the Removal of Asbestos-Containing Materials (ACM) from Buildings", American Industrial Hygiene Association Journal, June, 1991, pp. A324-328.

Kelse, J. W. and C. Sheldon Thompson, "The Regulatory and Mineralogical Definitions of Asbestos and Their Impact on Amphibole Dust Analysis", American Industrial Hygiene Association Journal, November 1989, pp. 613-622.

Webber, J. S., et al, "Quantitating Asbestos Content in Friable Bulk Samples: Development of a Stratified Point-Counting Method", American Industrial Hygiene Association Journal, August 1990, pp. 447-452.

WORKPLACE HEALTH PROTECTION PROGRAM ELEMENT CHECKLIST ON ASBESTOS

		Yes	No	Don't Know	N/A
1.	Are asbestos-containing materials (ACM) present within, or being used in the facility?	___	___	___	___
2.	Has a determination been made of where ACM is present in the facility?	___	___	___	___
3.	Is there a written operations and maintenance (O&M) program in effect which is designed to minimize contact with ACM and its disturbance?	___	___	___	___
4.	Is ACM inspected periodically?	___	___	___	___
5.	Are results of ACM inspections documented?	___	___	___	___
6.	Has a policy been established to eliminate the future use of ACM in the facility?	___	___	___	___
7.	Is there a written ACM control program?	___	___	___	___
8.	Has an individual or group been identified to be responsible for the development, implementation, and oversight of the facility's asbestos O&M and ACM handling program?	___	___	___	___
9.	Are ACM contractor bid proposals reviewed to ensure compliance with regulations, established procedures, accepted work practices, and company guidelines?	___	___	___	___
10.	Is a listing of acceptable ACM contractors maintained?	___	___	___	___
11.	Are contractor employees trained and approved by license (as required) to do asbestos work?	___	___	___	___

N/A - Not Applicable

WORKPLACE HEALTH PROTECTION PROGRAM ELEMENT CHECKLIST ON ASBESTOS

	Yes	No	Don't Know	N/A
12. Are requirements met with regard to notification of regulatory agencies when asbestos work is to be done?	___	___	___	___
13. Is appropriate protective equipment available for asbestos work:				
. Coveralls (type________)?	___	___	___	___
. Hair covering?	___	___	___	___
. Shoe covers?	___	___	___	___
. Gloves?	___	___	___	___
. Respiratory protection (type______)?	___	___	___	___
14. Have personnel who will be involved in handling ACM been provided:				
. Hazard awareness training?	___	___	___	___
. Medical examination?	___	___	___	___
. Respiratory protective equipment training?	___	___	___	___
. Procedure/work practice training?	___	___	___	___
. Information on the facility's workplace health protection/ safety practices and emergency procedures?	___	___	___	___
15. Have personnel involved in ACM work been entered in an asbestos medical surveillance program?	___	___	___	___
16. Have personnel who are potentially exposed to asbestos been fit tested for respirator use?	___	___	___	___
17. Is food consumption, drinking, and smoking forbidden in ACM abatement work areas?	___	___	___	___

N/A - Not Applicable

WORKPLACE HEALTH PROTECTION PROGRAM ELEMENT CHECKLIST ON ASBESTOS

		Yes	No	Don't Know	N/A
18.	Is there a practice of covering or removing furniture when there is a potential for its being contaminated by ACM during maintenance/asbestos abatement work?	___	___	___	___
19.	Are warning signs posted and barriers established at access ways to work areas where ACM may be disturbed?	___	___	___	___
20.	Are ACM control and containment procedures adequate:				
	. Wet methods used?	___	___	___	___
	. Enclosure used (when appropriate)?	___	___	___	___
	. Negative air used with enclosure (when appropriate)?	___	___	___	___
	. Glove bags used (when appropriate)?	___	___	___	___
	. HEPA filters changed as needed?	___	___	___	___
	. Segregated area established?	___	___	___	___
	. Limited access established?	___	___	___	___
	. Limited access maintained?	___	___	___	___
	. Drop cloths used (when appropriate)?	___	___	___	___
	. Housekeeping adequate?	___	___	___	___
	. Shower water filtered?	___	___	___	___
21.	Are abated equipment/surfaces adequately cleaned before an ACM substitute is applied:				
	. Definition established for what is clean?	___	___	___	___
	. Inspected before replacement applied?	___	___	___	___
	. No visible asbestos allowed?	___	___	___	___
	. Sealant applied to contain any possible remaining fibers?	___	___	___	___

N/A - Not Applicable

WORKPLACE HEALTH PROTECTION PROGRAM ELEMENT CHECKLIST ON ASBESTOS

	Yes	No	Don't Know	N/A
22. Does a facility individual (e.g., ACM coordinator) conduct oversight inspections of contractor work with ACM or at activities which can disturb ACM?	___	___	___	___
23. In the event an ACM containment enclosure is breached, does the contractor have a procedure for stopping work, correcting the breach, etc.?	___	___	___	___
24. Does contractor provide bulk sample information and air monitoring results to the facility ACM coordinator?	___	___	___	___
25. Do personnel performing ACM work use proper:				
. Coveralls (Type___________)?	___	___	___	___
. Hair covering?	___	___	___	___
. Shoe covers?	___	___	___	___
. Respiratory protection (Type__________)?	___	___	___	___
26. Have those who launder asbestos contaminated clothing been made aware of the presence of the asbestos contamination?	___	___	___	___
27. Are shower facilities available for personnel who may become contaminated with ACM?	___	___	___	___
28. Are disposal procedures adequate:				
. Properly labeled plastic bags used?	___	___	___	___
. Wastes retained in restricted area?	___	___	___	___
. Used HEPA filters effectively sealed, labeled and disposed?	___	___	___	___

N/A - Not Applicable

WORKPLACE HEALTH PROTECTION PROGRAM ELEMENT CHECKLIST ON ASBESTOS

	Yes	No	Don't Know	N/A
. Wastes disposed in an approved location?	___	___	___	___
. Contaminated protective clothing disposed as asbestos waste?	___	___	___	___
29. Is air monitoring routinely done during work with ACM:				
. Personal air samples?	___	___	___	___
. Area air samples:				
- Within work area?	___	___	___	___
- Contiguous areas?	___	___	___	___
- Clearance samples?	___	___	___	___
- After work is completed?	___	___	___	___
30. Are applicable records of asbestos work maintained:				
. Personal air monitoring results?	___	___	___	___
. Area air monitoring results?	___	___	___	___
. Clearance sampling results?	___	___	___	___
. Daily log of activities?	___	___	___	___
. Comments regarding contractor performance?	___	___	___	___
. Information regarding how much, who, when, and where disposal of ACM containing waste was made?	___	___	___	___
31. Are aggressive sampling procedures carried out for indoor clearance assessment?	___	___	___	___
32. Are workers who are exposed to asbestos in their work prohibited from wearing street clothes on the job?	___	___	___	___
33. Do workers who are exposed to asbestos (either facility or contractor) shower at the end of the workshift?	___	___	___	___

N/A - Not Applicable

WORKPLACE HEALTH PROTECTION PROGRAM ELEMENT CHECKLIST ON ASBESTOS

	Yes	No	Don't Know	N/A
34. Is the ACM work area inspected upon completion of asbestos abatement work to ensure no contamination exists and signs, plastic sheeting, etc. have been removed?	___	___	___	___
35. When practical, is asbestos work in inside areas (e.g., office buildings) limited to non-occupancy periods?	___	___	___	___
36. Is equipment that is used to drill/cut asbestos-containing material equipped with effective local exhaust ventilation?	___	___	___	___
37. Are disposable coveralls, gloves, and approved respiratory protection worn by personnel accessing drop ceiling spaces or crawl spaces where asbestos-containing material exists?	___	___	___	___
38. Is a spray-on sealant (non-air propelled) applied to friable ACM edges that are exposed in the course of work activities?	___	___	___	___
39. Are tools cleaned after use where asbestos contamination may occur?	___	___	___	___
40. Where practical, is a decontamination facility (negative air/dirty side/shower/clean side) established for personnel performing asbestos abatement work?	___	___	___	___
41. Is bagged asbestos waste placed in a second, clean, properly labeled plastic bag that is effectively closed (taped shut) before disposal?	___	___	___	___

N/A - Not Applicable

WORKPLACE HEALTH PROTECTION PROGRAM ELEMENT CHECKLIST ON ASBESTOS

		Yes	No	Don't Know	N/A
42.	Is a HEPA vacuum unit required at ACM work locations for maintaining acceptable housekeeping?	___	___	___	___
43.	Is a list of emergency phone numbers available at the ACM abatement work area?	___	___	___	___
44.	Have written procedures been established for ACM abatement personnel to follow if an emergency develops during work?	___	___	___	___
45.	Has the potential for excessive heat stress been addressed during ACM abatement work?	___	___	___	___
46.	Is a supply of cool potable drinking water provided in the vicinity of asbestos abatement work locations?	___	___	___	___
47.	Are access/egress procedures for the asbestos abatement work area acceptable?	___	___	___	___

N/A - Not Applicable

WORKPLACE HEALTH PROTECTION PROGRAM ELEMENT
ON
COMPRESSED GASES

WORKPLACE HEALTH PROTECTION PROGRAM ELEMENT ON COMPRESSED GASES

OBJECTIVE

To provide information on the health protection considerations in the handling and use of compressed gases.

BACKGROUND

Gases are substances which exist in the gaseous state at normal temperature and pressure [approximately 70°F (21°C) and 14.7 psia (101 kPa)]. Compressed gases are those which remain in the gaseous state in a containment vessel, whereas liquified gases are those that, at normal temperature inside a closed container, exist partly in the liquid state and partly in the gaseous state. Cryogenic gases are liquified gases that must be maintained at far below normal temperature to keep them in the liquified state.

Gases are used for many purposes in petroleum/petrochemical operations. Typical applications include carbon dioxide and Halon as fire suppressants; nitrogen for inerting equipment; helium, hydrogen, nitrogen, ammonia, nitrous oxide, etc. for laboratory instruments and process analyzer operation; chlorine and hydrogen sulfide for catalyst activation; chlorine for water treatment; and gases for welding operations. There is a potential for personnel to be exposed to these gases during use, when connecting regulators/transfer lines, as a result of leakage, from discharges during emergencies, and when disconnecting cylinders. The potential hazards presented to personnel include corrosive effects, eye and respiratory tract irritation, asphyxiation, systemic effects, and others.

Compressed gases are received in a variety of forms, including rail car and tank truck, as well as large and small cylinders. A good practice is to limit the storage of gases to present needs.

CONSIDERATIONS

In order to reduce the potential for exposure to compressed gases during handling and use, it is essential that personnel be aware of the associated hazards, effective procedures/work practices be established and followed, and appropriate protective equipment be made available and used by personnel.

Compressed gas cylinders must be properly identified, handled and stored. Storage should be in well ventilated areas away from sources of heat. Pressure regulators must be the type specified for both the cylinder and the gas. No jury-rigging should be permitted. Color coding of cylinders should not be relied upon solely as a gas content identification method. Cylinders should be dated when received and used in sequence relative to the date of receipt. Records should be maintained of the gases that are stored and in use, as well as their location. This information should be available in the event of an emergency.

Gas transfer line material should be compatible with the gas being handled. Lines should be cleaned before introducing the gas. Cleaning generally involves use of a non-flammable solvent being pumped through the line, followed by drying of the line with nitrogen. In addition, leak testing should be performed. This can be done while nitrogen gas is being used in the drying procedure. A pre-operational check should be carried out to ensure the storage location, regulator, and transfer system are appropriate. The compressed gas storage area should be away from building air intakes. Oxidants should not be stored with flammables. If gases are to be used in a tank, or other confined space, the cylinder should be secured outside. Gas cylinders should not be brought into such locations. In most situations, it is inadvisable to store and use gas cylinders inside a building. If this is necessary, the smallest available cylinder should be used. If a cylinder of toxic gas is to be used inside, it should be kept in a mechanically exhaust ventilated hood or gas cabinet. In addition, it would be advisable to install a continuous monitor in inside locations where highly toxic gases are in use.

Caps should be kept on cylinders when not in use and while being moved. Personnel making cylinder connections should be properly trained, and be provided and use appropriate protective equipment. Excessive force should not be applied when opening cylinder outlet valves. If a valve is stuck, the cylinder should be returned to the supplier. An automatic gas shut-off valve should be incorporated in a toxic gas supply system to prevent excessive contaminant release if the gas flow exceeds a pre-set rate. Gas supply systems should be checked periodically for leakage. Personnel should know the location of gas shut-off valves, particularly those handling toxic or flammable gases.

REFERENCES

"Handbook Of Compressed Gases", Compressed Gas Association, 2nd Edition (New York: Van Nostrand Reinhold, 1981).

"Guide For Handling of Compressed Gases" (East Rutherford, NJ: Matheson Co., 1982).

"Matheson Gas Data Book" (East Rutherford, NJ: Matheson Co., 1980).

Walls, W. L., "Storage of Gases", Fire Protection Handbook, 16th Edition, Chapter 5 (Quincy, MA: National Fire Protection Association, 1986), pp. 39-58.

WORKPLACE HEALTH PROTECTION PROGRAM ELEMENT CHECKLIST ON COMPRESSED GASES

		Yes	No	Don't Know	N/A
1.	Are compressed gas cylinders, transfer lines and regulators identified?	___	___	___	___
2.	Are gas cylinders dated upon receipt?	___	___	___	___
3.	Are compressed gas cylinders properly stored and supported?	___	___	___	___
4.	Is the gas cylinder storage area located a sufficient distance from air inlets (i.e., more than 75 feet)?	___	___	___	___
5.	Is the gas storage area housekeeping satisfactory?	___	___	___	___
6.	Where appropriate, is the gas cylinder storage area provided with a separate ventilation system?	___	___	___	___
7.	Is a wheeled cart available and used to move large gas cylinders?	___	___	___	___
8.	Are the regulators and transfer lines proper for the cylinders and gases?	___	___	___	___
9.	Are gas transfer lines cleaned before they are placed in service?	___	___	___	___
10	Do personnel using and/or working with gas systems know how and where to shut off the gas supply?	___	___	___	___
11.	Is there an automatic valve in highly toxic gas supply systems to shut off the supply if there is an excess flow?	___	___	___	___
12.	Are toxic gas cylinder connections and associated transfer lines periodically checked for leakage?	___	___	___	___

N/A - Not Applicable

WORKPLACE HEALTH PROTECTION PROGRAM ELEMENT CHECKLIST ON COMPRESSED GASES

		Yes	No	Don't Know	N/A
13.	Is there a provision to prevent the installation of gas transfer lines above occupied areas (other than a laboratory)?	___	___	___	___
14.	Are oxidant gases adequately separated from flammable gases?	___	___	___	___
15.	Are personnel who may be exposed to hazardous gases trained with respect to handling procedures, hazards, toxic effects, etc.?	___	___	___	___
16.	Is appropriate protective equipment available and used by personnel handling or using compressed/liquified/cryogenic gases?	___	___	___	___
17.	Are emergency response personnel informed of the presence and location of hazardous gases in the facility and of their potential adverse health effects?	___	___	___	___
18.	Are gas cylinders stored and used in well ventilated areas away from sources of heat and ignition?	___	___	___	___
19.	Are compressed gas cylinders not permitted to be stored or used in laboratories or other inside locations?	___	___	___	___
20.	Are gas cylinders, if used within the laboratory, limited by size and quantity?	___	___	___	___
21.	Are caps kept on cylinders when not in use and when being moved?	___	___	___	___
22.	Are cylinders located away from air intakes, loading docks, and spark producing equipment?	___	___	___	___

N/A - Not Applicable

WORKPLACE HEALTH PROTECTION PROGRAM ELEMENT CHECKLIST ON COMPRESSED GASES

		Yes	No	Don't Know	N/A
23.	Is there a procedure to test the air before re-entry is permitted into an area into which CO_2 or Halon has been discharged?	___	___	___	___
24.	Is available monitoring equipment adequate for testing air quality in an area where CO_2 or Halon has been discharged?	___	___	___	___
25.	Is there adequate warning to evacuate an occupied area before a CO_2 or Halon system is discharged (re: noise, toxicity, oxygen depletion concerns)?	___	___	___	___
26.	Are continuous monitors operated at inside locations where highly toxic compressed gases are in use?	___	___	___	___
27.	Are continuous monitors periodically checked to ensure proper operation and calibration?	___	___	___	___
28.	Are operation and calibration records of continuous monitors adequately maintained?	___	___	___	___
29.	Is there a practice to maintain a slight pressure in a cylinder when disconnecting if for replacement?	___	___	___	___
30.	Is a three tag identification system used for gas cylinders to indicate full, in-service, or empty?	___	___	___	___
31.	Are empty compressed gas cylinders segregated from full ones?	___	___	___	___

N/A - Not Applicable

WORKPLACE HEALTH PROTECTION PROGRAM ELEMENT CHECKLIST
ON
COMPRESSED GASES

	Yes	No	Don't Know	N/A
32. Are compressed gas cylinders kept outside a confined/enclosed space when the gas is being used inside?	___	___	___	___
33. Are Material Safety Data Sheets available on each compressed gas being stored or used?	___	___	___	___
34. Is there a procedure to check compressed breathing air cylinders to assure the contents meet breathing air specifications?	___	___	___	___
35. Are compression fittings on gas transfer systems all of the same manufacturer?	___	___	___	___

N/A - Not Applicable

WORKPLACE HEALTH PROTECTION PROGRAM ELEMENT
ON
SOUR STREAMS

WORKPLACE HEALTH PROTECTION PROGRAM ELEMENT ON SOUR STREAMS

OBJECTIVE

To provide information on the hazards of hydrogen sulfide gas and procedures/work practices that should be followed in areas/operations where there is a potential for exposure to this contaminant.

BACKGROUND

There may be a significant potential for personnel to be exposed to hydrogen sulfide gas in some operations carried out in the petroleum and petrochemical industries. This gas can be released, along with other gases and vapors, during well drilling, production of gas and crude oil, in the transportation of gas or crude by pipeline or ship, as well as in refining operations (desulfurization, sulfur recovery, etc.) and at marketing facilities (asphalt, and some fuel oil storage tanks).

Hydrogen sulfide is a colorless, flammable, highly toxic gas with a characteristic odor. Although this gas can be detected by its odor at very low concentrations, odor should not be relied upon as a means of detection or hazard assessment. At higher concentrations hydrogen sulfide paralyzes the sense of smell.

Sour is a term that is applied to the feed, intermediate, or product streams which contain malodorous sulfur compounds. In the petroleum industry the term sour is applied to materials which contain sulfur or sulfur compounds. However, there is no consistent agreement as to the concentration of sulfur or sulfur compounds that give rise to the classification of a stream or material as sour. For this discussion, materials which present a potential for release of hydrogen sulfide that can result in a concentration in air greater than 300 parts per million (ppm) are considered sour. This concentration is referred to as the IDLH (Immediately Dangerous to Life or Health) level for hydrogen sulfide. It is the maximum concentration from which one could escape within 30 minutes without any escape-impairing symptoms or irreversible adverse health effect.

The potential for an H_2S release, along with the concentration inside process equipment or piping, has been used as a basis for determining how hazardous an area could be if a release occurred. For example, if the H_2S concentration in piping or equipment is above 0.5 percent by volume and there is a potential for a release, then the area could be considered one of high hazard potential. If the concentration in a gas stream or vapor space is below 300 ppm (0.03%) by volume, and there is a potential for a release, then the area could be classified as one of low hazard potential. For those areas where there is a potential for release and the concentration of H_2S in the stream or vapor space is between 0.03% and 0.5% by volume the area could be considered one of moderate H_2S hazard potential.

CONSIDERATIONS

Since hydrogen sulfide is an acutely acting, highly toxic material to which personnel may be exposed in a number of petroleum and petrochemical operations, it is essential that those who may be exposed to this gas be aware of its hazardous properties. All potentially exposed employees should be provided periodic hazard awareness training on H_2S to maintain their alertness to its hazards, what they can do to reduce their exposure potential, and the action to take if an individual is overcome by hydrogen sulfide.

Areas and operations where there is a potential for exposure to hydrogen sulfide should be identified by appropriate signs. A classification scheme, based on the concentration of H_2S in process equipment or vessels and the potential for a gas release to occur, can be applied to determine the area/operation hazard potential. Based on the results of these assessments, decisions can be made with regard to the need for signs, personnel training, respiratory protection, continuous monitors, etc.

Personnel who may be exposed to hydrogen sulfide must be provided respirator and pulmonary resuscitation training. Designated personnel should be provided training in cardio-pulmonary resuscitation (CPR), where warranted.

Procedures/work practices must be developed and implemented to reduce the potential for exposure to hydrogen sulfide gas. For example, opening and draining equipment, installing blinds, or sampling and gauging crude, gas or product which contains hydrogen sulfide can result in exposure to this toxic gas. Use of appropriate respiratory protection is necessary for some of these tasks. For some situations the buddy system must be adhered to. This involves two persons, properly equipped, carrying out a task such that the second person provides back-up for the other who performs the work.

Purging and flushing of equipment or lines which contain H_2S must be effective for removal of this gas before the equipment is opened. If there is any question at all about H_2S removal effectiveness, then appropriate respiratory protection should be used when opening the equipment. Sampling and gauging of tanks containing sour crude or sour product, as well as going onto the floating roof of a tank that contains a sour material, requires use of respiratory protection and adherence with the buddy system. There are numerous other tasks that are carried out which require use of respiratory protection and a back-up person. These must be identified and appropriate procedures/work practices implemented.

Emergency respiratory protective equipment (positive pressure self-contained breathing units with full facepiece) should be located in potential H_2S contaminated areas/units. This equipment, which may best be located away from the center of the potentially contaminated area, is to be used only for emergency situations and should be periodically checked (monthly) to ensure each respirator is in operating condition.

Personal and portable H_2S monitoring equipment should be available to process and maintenance personnel to enable them to check for the presence of H_2S contamination, where appropriate. This practice should not replace the permit procedure in which gas testing is performed to determine if gas concentrations are satisfactory for the work to be performed and specifying the protective equipment needs for use while performing the task.

For potential high hazard areas it would be appropriate to install continuous H_2S monitors with the sensor about 2 feet above grade. Their purpose is to provide prompt detection of a gas release and early warning of the presence of H_2S contamination so that measures can be taken to prevent excessive exposure of personnel and reduce area contamination. These instruments should be positioned near locations/equipment most likely to be a source of H_2S release. This may include locations in the vicinity of compressor seals, amine scrubbers, sour water strippers, in the sulfur recovery facility, tail gas unit, and others. At least two persons, appropriately equipped (positive pressure self-contained breathing equipment with full facepiece and an H_2S monitoring device), should respond to an H_2S alarm situation. They should approach the source of H_2S release from an upwind direction and not remove the respiratory protective equipment until test results indicate the H_2S level in the affected area is acceptable. Canister or cartridge-type respirators are not to be used. Only self-contained breathing equipment should be used.

Contractors should receive hazard awareness training before they begin work in potential H_2S contaminated areas. Facility personnel should check on contractor performance frequently to ensure their adherence with permit requirements and accepted procedures/work practices.

Contractor personnel should be aware of emergency signals and procedures to follow if an emergency arises.

It is advisable to provide personnel working in potentially high H_2S areas with personal alarms. These will provide a warning of a potentially hazardous exposure situation if the concentration of H_2S in the breathing zone exceeds a preset alarm level (generally 10 or 20 ppm).

Where appropriate, wind socks should be placed at several highly visible elevated positions throughout the facility so that egress can be made in an appropriate direction in the event of an H_2S release. Personnel should be trained on how to interpret wind sock orientation for selecting an evacuation route.

REFERENCES

"Criteria for a Recommended Standard-Occupational Exposure to Hydrogen Sulfide" [Cincinnati, OH: U. S. Department of Health, Education and Welfare, DHEW Publication No. (NIOSH) 77-158, 1977].

Ludovich, D., "The Hydrogen Sulfide Technical Manual" (Lafayette, LA: Safety Technology & Oilfield Protectors, 1981).

Halley, P.D., "Hazards of Hydrogen Sulfide", Paper presented at the National Safety Congress (Chicago, IL: 1966).

Poda, G. A., "Hydrogen Sulfide Can Be Handled Safely", Arch. Environmental Health, Vol. 12, June 1966.

Slack, D. J., "Hydrogen Sulfide in Residual Fuel Oil and Storage Tank Vapor Space", American Industrial Hygiene Association Journal, April 1988, pp. 205-206.

"Hydrogen Sulfide Manual" (Paris, France: Schlumberger, 1982).

Lagas, J. A., "Stop Emissions from Liquid Sulfur", Hydrocarbon Processing, October 1982, pp. 85-89.

WORKPLACE HEALTH PROTECTION PROGRAM ELEMENT CHECKLIST ON SOUR STREAMS

		Yes	No	Don't Know	N/A
1.	Have areas/operations been identified/classified with respect to H_2S exposure potential?	___	___	___	___
2.	Have appropriate procedures/work practices been developed in recognition of the H_2S exposure potential?	___	___	___	___
3.	Have personnel who may be exposed to H_2S been provided:				
	. Training with respect to its hazards?	___	___	___	___
	. Respiratory protective equipment training?	___	___	___	___
	. Pulmonary resuscitation or CPR training?	___	___	___	___
	. Emergency response training?	___	___	___	___
	. Medical surveillance?	___	___	___	___
4.	Are warning signs posted in potential H_2S exposure areas?	___	___	___	___
5.	Are there warning signs at stairways/ladders on tanks containing a source of H_2S?	___	___	___	___
6.	Is adequate emergency respiratory protective equipment effectively located in potential H_2S contaminated areas?	___	___	___	___
7.	Are personnel who work in potential H_2S contaminated areas provided personal alarms?	___	___	___	___
8.	Are personnel properly trained to respond to an H_2S release?	___	___	___	___
9.	Are there calibrated, portable H_2S instruments available for area and confined space monitoring and source identification?	___	___	___	___

N/A - Not Applicable

WORKPLACE HEALTH PROTECTION PROGRAM ELEMENT CHECKLIST ON SOUR STREAMS

	Question	Yes	No	Don't Know	N/A
10.	Are continuous H_2S monitors in operation in high potential H_2S release areas?	___	___	___	___
11.	Are continuous H_2S area monitors:				
	. Periodically response checked?	___	___	___	___
	. Periodically calibrated?	___	___	___	___
	. Records retained of maintenance and calibration?	___	___	___	___
12.	Do continuous H_2S monitors alarm both locally and in an occupied control center?	___	___	___	___
13.	Is there always a personnel response to a continuous H_2S monitor alarm?	___	___	___	___
14.	Are continuous H_2S monitors considered critical equipment?	___	___	___	___
15.	Is the buddy system required for:				
	. Responding to H_2S emergencies?	___	___	___	___
	. Investigating H_2S releases?	___	___	___	___
	. Gauging sour tanks?	___	___	___	___
	. Sampling sour tanks?	___	___	___	___
	. Other tasks where exposure to H_2S may occur?	___	___	___	___
16.	Is H_2S containing piping/ equipment identified?	___	___	___	___
17.	Are H_2S gas tests conducted before opening lines/equipment that contained a sour material?	___	___	___	___
18.	Where appropriate, are self-contained breathing units worn when opening lines/equipment that contained H_2S?	___	___	___	___
19.	Are periodic reviews conducted to minimize the exposure potential at operations/jobs at which H_2S contamination may exist?	___	___	___	___

N/A - Not Applicable

WORKPLACE HEALTH PROTECTION PROGRAM ELEMENT CHECKLIST ON SOUR STREAMS

	Yes	No	Don't Know	N/A
20. Are personnel aware that cartridge and canisters type respiratory protection is unacceptable for potential H_2S exposure situations?	___	___	___	___
21. Are there observable wind direction indicators (e.g., wind socks) throughout the facility or on vessels transporting sour product?	___	___	___	___
22. Are facility personnel informed of appropriate evacuation routes and assembly areas?	___	___	___	___
23. Have contractor personnel working in potential H_2S contaminated areas been:				
. Provided H_2S hazard training?	___	___	___	___
. Made aware of the facility's accepted procedures/work practices with respect to H_2S?	___	___	___	___
. Provided respirator training?	___	___	___	___
. Made aware of emergency response procedures (e.g., alarm signals, evacuation routes, assembly areas)?	___	___	___	___
24. Is there a written contingency plan for an H_2S emergency?	___	___	___	___
25. During start-up of high H_2S source units (e.g., sulfur recovery facility):				
. Are flanges, valve packings, instrument connections, etc. checked for leaking gas?	___	___	___	___
. Do personnel wear an SCBA unit during unit/equipment start-up leak checks?	___	___	___	___

N/A - Not Applicable

WORKPLACE HEALTH PROTECTION PROGRAM ELEMENT CHECKLIST ON SOUR STREAMS

		Yes	No	Don't Know	N/A
26.	Are acid cleaning procedures reviewed before being carried out to ensure H_2S formation potential is negligible?	___	___	___	___
27.	Are seals on equipment (e.g., pumps, compressors, etc.) in sour service adequately maintained to minimize H_2S leakage?	___	___	___	___
28.	Are only those personnel in the respirator medical surveillance program permitted to work in potential H_2S contaminated areas?	___	___	___	___
29.	Are precautions taken to reduce H_2S exposure potential when draining sour water from tanks, separators, etc.?	___	___	___	___
30.	Is the water draw-off from sour crude tanks into a closed system?	___	___	___	___
31.	Are vents which may release H_2S containing streams high enough to prevent significant contamination of work locations?	___	___	___	___
32.	Are vents and safety valves on H_2S containing equipment directed to a flare?	___	___	___	___
33.	Do personnel working on elevated platforms carry respiratory protection (escape type) with them when working in potential H_2S contaminated areas?	___	___	___	___
34.	Is the sulfur storage pit and/or sulfur storage tank purge system checked periodically to ensure effective purging of off-gases?	___	___	___	___

N/A - Not Applicable

**WORKPLACE HEALTH PROTECTION PROGRAM ELEMENT CHECKLIST
ON
SOUR STREAMS**

	Yes	No	Don't Know	N/A
35. Is the liquid sulfur storage pit/ tank purge gas directed to a flare or sulfur plant incinerator?	___	___	___	___
36. Do personnel working around molten sulfur wear appropriate personal protective equipment (eye, face, hand, skin, and respiratory protection, as appropriate)?	___	___	___	___
37. Is the liquid sulfur loading arm (i.e., for tank trucks and rail cars) provided with a "dead-man" switch?	___	___	___	___
38. Are workplace health protection reviews conducted when there is a change in processing or handling of materials that contain hydrogen sulfide?	___	___	___	___

N/A - Not Applicable

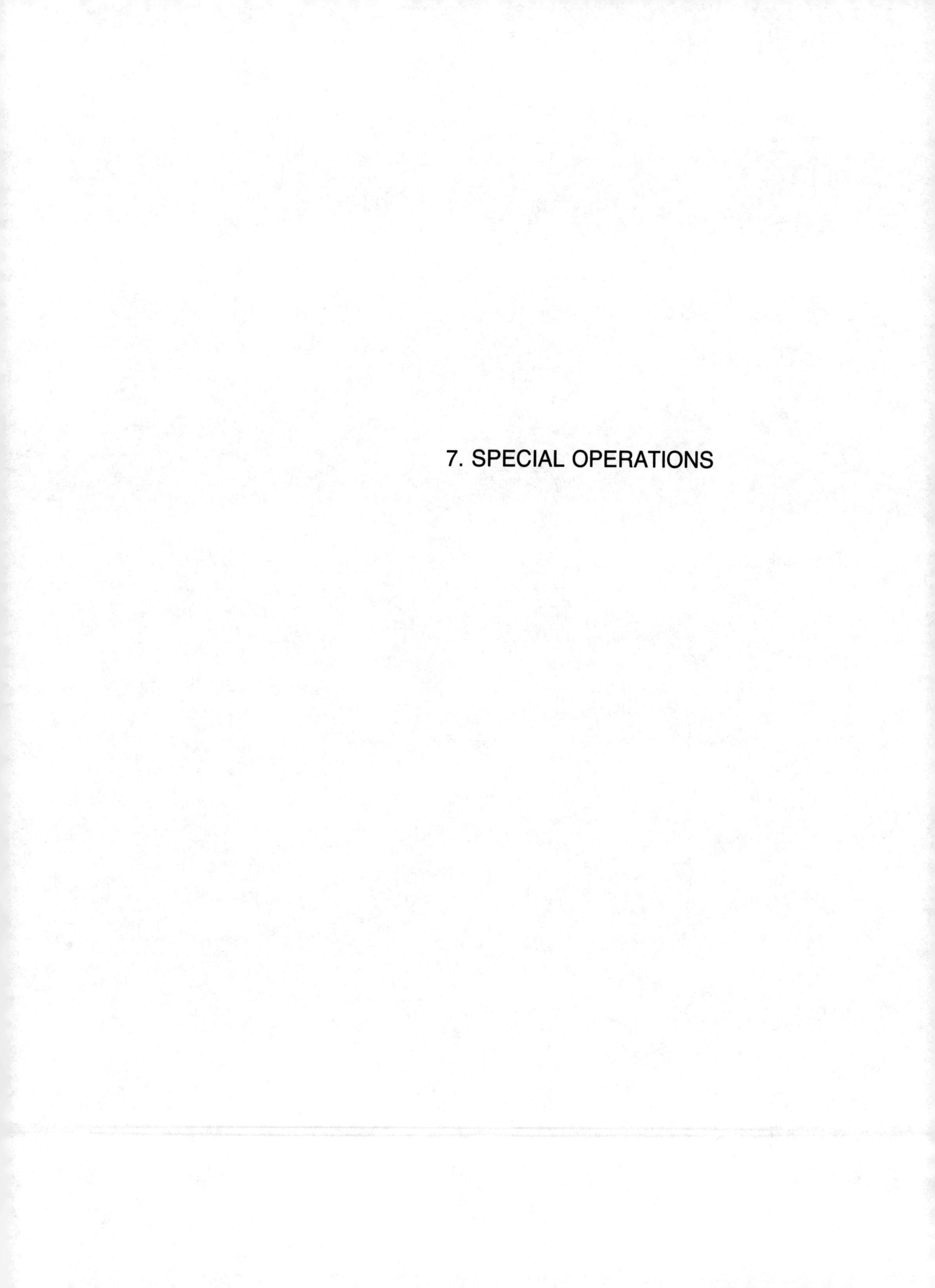

7. SPECIAL OPERATIONS

WORKPLACE HEALTH PROTECTION PROGRAM ELEMENT
ON
ABRASIVE CLEANING

WORKPLACE HEALTH PROTECTION PROGRAM ELEMENT ON ABRASIVE CLEANING

OBJECTIVE

To provide information on abrasive cleaning operations and methods for reducing the potential health hazards associated with this work.

BACKGROUND

Abrasive cleaning is a generic term applied to cleaning operations in which a solid abrasive agent is directed at a surface by air pressure, centrifugal force, or hydraulically to remove rust, paint, soil, or roughness. This procedure is frequently carried out to prepare metal parts, structures, tanks, or other equipment for painting or other surface coating.

Abrasive cleaning systems range from small cabinets with gloveports to large equipment consisting of a compressor, abrasive hopper, hoseline with a blast nozzle, and an abrasive reclaimer. In most operations, an individual directs the abrasive at the surface being cleaned. Some operations, however, are automated.

Materials employed for abrasive cleaning are classified according to size, shape and hardness. They are selected based on the characteristics of the surface to be cleaned and the desired finish. Materials used include steel shot, glass beads, slag, silicon carbide, beach sand, ground walnut shells, rice hulls, aluminum oxide and others. Use of abrasives containing crystalline silica are not recommended due to their potential health effects.

Personnel can be exposed to airborne contaminants and physical agents during abrasive cleaning operations. These include abrasive dust, noise, paint dust, heat, and others. The surface coating materials being removed may contain lead, or chromate pigment. The consequence of excessive exposure to these may be serious due to the toxicity of the compounds present. Exposure controls, in the form of personal protective equipment, exhaust ventilation, enclosures, wet procedures and others are important for reducing the health hazards associated with abrasive cleaning operations.

CONSIDERATIONS

Personnel involved in abrasive cleaning operations must be trained in the use of equipment (abrasive cleaning and personal protection), made aware of the potential health hazards associated with the work, and know when to use personal protective equipment to minimize their exposure to the hazardous agents associated with this work.

Access to abrasive cleaning operating areas should be restricted to minimize the exposure of non-operating personnel to dust and noise. Rope barriers and signs should be used to accomplish this. Appropriate protective equipment is to be worn by personnel if they enter the restricted area.

Sheeting (tarpaulin, plastic, plywood) or other type enclosure should be used, when practical, to prevent the spread of abrasive/pigment/metal dust. Where feasible, wet systems should be employed. Exhaust ventilation should be provided, where practical, to control airborne particulate contamination.

The least hazardous abrasive material should be used. Those which contain crystalline silica (quartz, cristobalite or tridymite) should not be used.

The composition of the material to be removed in the cleaning operation should be determined. Coatings containing lead, zinc or chromate pigment can present a risk to personnel if the procedures, work practices and/or exposure controls are not adequate. Medical surveillance may be appropriate for those operations where there is a potential for excessive exposure to lead, silica or other toxic material.

Personnel involved in abrasive cleaning operations must be provided and required to wear appropriate personal protective equipment. Items needed for operations in the open air include an approved abrasive blast helmet with supplied breathing air; eye protection; gloves; as well as foot, body, and hearing protection. Similar protective equipment would be appropriate for work inside tanks, except the added concerns associated with work in a confined space must also be addressed. Personnel who enter an abrasive cleaning work area (i.e., within the rope barrier) should wear appropriate eye protection, as well as hearing and respiratory protection, if necessary.

Air supplied to respiratory protective equipment must meet breathing air specifications. A breathing air compressor is recommended. If the compressor used to supply air is not a breathing air type unit, the air supply system should be equipped with high temperature and carbon monoxide alarms. The compressor should be located where it is unlikely that combustion exhaust from the compressor motor, or contaminated ambient air will enter the system. Air supplied to abrasive cleaning personnel should be tested periodically to ensure its quality is acceptable.

Abrasive cleaning personnel must be provided training on procedures/work practices to follow, how to use available protective equipment and information on the hazards associated with the work.

The abrasive cleaning equipment and the work area should be checked periodically to ensure the equipment is operating properly, there is no leakage of abrasive from the hose, and personnel within the restricted area are wearing appropriate protective equipment.

The abrasive discharge nozzle should be equipped with a control valve of the "deadman type" so that abrasive flow will be stopped if the actuator is not manually depressed.

Abrasive cleaning personnel are exposed to noise from the compressor operation, equipment vibration, impact of abrasive striking the surface being cleaned, air pressure drop at the nozzle, and from breathing air supplied to the abrasive blast helmet. Noise exposure of the operator and others within the work area can be excessive and exposures need to be controlled. Acoustically enclosed compressors are available and should be used. Mufflers on internal combustion engine exhausts should be effective for reducing noise emissions. Air supply to the abrasive blast helmet should be adequate for breathing purposes but not such that noise energy associated with its discharge in the helmet is excessive.

Asbestos fibers may be generated during abrasive cleaning of structures or surfaces that had previously been covered with asbestos-containing material. Tower trays to be cleaned may have surface contamination from catalysts (cobalt, nickel, molybdenum, etc.) or other hazardous substance. Information, such as the foregoing, must be developed on all aspects of the abrasive cleaning task so that the potential risks associated with the work can be effectively identified and exposure controls implemented before it is initiated.

REFERENCES

"Industrial Ventilation - A Manual of Recommended Practice", 20th Edition (Cincinnati, OH: American Conference of Governmental Industrial Hygienists, 1988).

"American National Standard for Exhaust Systems - Abrasive Blasting Operations", ANSI/ASC Z9.4 (New York: American National Standards Institute, 1985).

"Safe Maintenance Practices in Refineries", API Publication 2007, Second Edition (Washington, D. C.: American Petroleum Institute, February 1983).

Burgess, W. A., "Potential Exposures in Industry - Their Recognition and Control", Patty's Industrial Hygiene and Toxicology, Vol. 1, General Principles, Third Edition (New York: John Wiley & Sons, Inc., 1978), pp. 1150-1151.

WORKPLACE HEALTH PROTECTION PROGRAM ELEMENT CHECKLIST ON ABRASIVE CLEANING

		Yes	No	Don't Know	N/A
1.	Has a health hazard review been carried out of the abrasive cleaning procedures/work practices?	___	___	___	___
2.	Are Material Safety Data Sheets available on the abrasive cleaning materials?	___	___	___	___
3.	Are crystalline silica-containing abrasives disallowed for use in abrasive cleaning operations?	___	___	___	___
4.	Are personnel engaged in abrasive cleaning operations provided training on procedures/work practices and of the potential health hazards associated with the work?	___	___	___	___
5.	Is access to abrasive cleaning areas effectively restricted to minimize exposure of non-operating personnel to contaminants?	___	___	___	___
6.	Is there a procedure in place to ensure that work practices, area access restrictions, equipment, personal protection, etc., are effective for reducing exposure of personnel to airborne contaminants and physical agents associated with abrasive cleaning operations?	___	___	___	___
7.	Has personal monitoring been conducted to assess worker exposure to airborne contaminants and physical agents during abrasive cleaning operations?	___	___	___	___

N/A - Not Applicable

WORKPLACE HEALTH PROTECTION PROGRAM ELEMENT CHECKLIST ON ABRASIVE CLEANING

		Yes	No	Don't Know	N/A
8.	Are personnel who are engaged in abrasive cleaning provided:				
	. Periodic medical examinations?	___	___	___	___
	. Pulmonary function tests?	___	___	___	___
	. Audiometric tests?	___	___	___	___
	. Medical surveillance for respirator use and specific contaminants (e.g., lead, vanadium, etc.)?	___	___	___	___
9.	Are at least two people assigned to abrasive cleaning tasks in tanks, towers, or cleaning rooms?	___	___	___	___
10.	Is it a practice to identify the composition of surface coatings to be removed, as well as other potential hazards associated with an abrasive cleaning operation?	___	___	___	___
11.	Are potential abrasive cleaning contractors informed of the surface coating material composition before bid submission?	___	___	___	___
12.	During abrasive cleaning work, is the proper personal protective equipment provided and worn by:				
	. Operator?	___	___	___	___
	. Helper?	___	___	___	___
	. Non-operating personnel entering the work area?	___	___	___	___
13.	Is personal protective equipment:				
	. Clean?	___	___	___	___
	. Properly stored?	___	___	___	___
	. Properly maintained?	___	___	___	___
14.	Have abrasive cleaning personnel been trained in the use of respiratory protective equipment?	___	___	___	___
15.	Is acceptable breathing air supplied to abrasive cleaning personnel?	___	___	___	___

N/A - Not Applicable

WORKPLACE HEALTH PROTECTION PROGRAM ELEMENT CHECKLIST ON ABRASIVE CLEANING

		Yes	No	Don't Know	N/A
16.	Is the breathing air compressor located in an area where it is unlikely that combustion exhaust or other contaminant will enter the system?	___	___	___	___
17.	Does the breathing air system have:				
	. Pressure regulation?	___	___	___	___
	. Oil mist and particulate filter?	___	___	___	___
	. High temperature alarm (when appropriate)?	___	___	___	___
	. Combination filter bed for particulates, CO, etc.?	___	___	___	___
	. Adequate air supplied to personnel?	___	___	___	___
18.	Are breathing air supply systems that are equipped with combination filters (e.g., to remove particulates, hydrocarbon vapors, carbon monoxide, etc.) adequately maintained to ensure effectiveness?	___	___	___	___
19.	Is the abrasive blast cleaning system inspected each day it is operated to ensure abrasive is not leaking?	___	___	___	___
20.	Is the abrasive cleaning nozzle a "deadman" type?	___	___	___	___
21.	Is housekeeping in the abrasive cleaning work area acceptable?	___	___	___	___
22.	Are measures taken to reduce the dispersion of airborne particulate contamination during abrasive cleaning operations?	___	___	___	___

N/A - Not Applicable

WORKPLACE HEALTH PROTECTION PROGRAM ELEMENT CHECKLIST ON ABRASIVE CLEANING

	Yes	No	Don't Know	N/A
23. Have wet methods been considered for reducing dust generation at abrasive cleaning operations?	___	___	___	___
24. Has paint stripping been considered as an alternative to abrasive cleaning?	___	___	___	___
25. Is the dust collection system for abrasive cleaning cabinets/rooms/enclosures operating properly (e.g., fan operating per design, bags not leaking, waste hoppers emptied periodically, etc.)?	___	___	___	___
26. Are abrasive cleaning exhaust ventilation systems evaluated periodically?	___	___	___	___
27. Do abrasive cleaning exhaust ventilation systems (where available) meet design specifications?	___	___	___	___
28. Is the abrasive cleaning work area adequately illuminated?	___	___	___	___
29. Are air compressors used in abrasive cleaning operations equipped with noise controls?	___	___	___	___
30. Are measures taken to reduce noise generation due to the vibration associated with abrasive cleaning operations?	___	___	___	___
31. Is hearing protection used by abrasive blast cleaning equipment operators?	___	___	___	___
32. Are waste abrasives and particulates from the cleaning operation properly disposed?	___	___	___	___

N/A - Not Applicable

WORKPLACE HEALTH PROTECTION PROGRAM ELEMENT CHECKLIST ON ABRASIVE CLEANING

	Yes	No	Don't Know	N/A
33. Are sanitation facilities adequate at abrasive cleaning operating areas to enable personnel to practice good personal hygiene?	___	___	___	___
34. Are eating facilities adequate for abrasive cleaning personnel?	___	___	___	___
35. Do abrasive cleaning personnel wash before eating?	___	___	___	___
36. Are personnel who are potentially exposed to lead provided:				
. Hazard awareness training (re: lead)?	___	___	___	___
. Medical examination (i.e., included in the lead surveillance program)?	___	___	___	___

N/A - Not Applicable

WORKPLACE HEALTH PROTECTION PROGRAM ELEMENT
ON
BULK PRODUCT LOADING

WORKPLACE HEALTH PROTECTION PROGRAM ELEMENT
ON
BULK PRODUCT LOADING

OBJECTIVE

To provide information to enable a review of product loading procedures/work practices so a judgment can be made of their effectiveness for minimizing exposure to the materials being handled.

BACKGROUND

Bulk liquid and gaseous materials are transferred from petroleum and petrochemical facilities to customers by tanktruck, railcar, or marine vessel (barge, tanker). Loading of these products is accomplished by transport personnel or facility employees.

Individuals involved in the loading operation may be exposed to the products being transferred. Exposure can be to the vapor emitted from the liquid product or to the liquid itself, as well as to a liquified gas or the gas itself. Exposures can occur when transfer equipment (fill pipe, hose, loading arm, etc.) is introduced into a tank, or connected to a manifold; during the filling procedure; when sampling the product; while removing or disconnecting transfer equipment; if a spill occurs; and during other phases of the operation.

Materials that are handled may be irritants (e.g., ammonia), corrosives (e.g., spent acids, spent caustic), systemic toxicants (e.g., benzene, 1,3-butadiene, etc.), asphyxiants (e.g., butane, ethylene, etc.), or possess other toxic properties. Exposure to these should be controlled so that adverse health effects do not arise.

The exposure potential presented to personnel is related to the amount of material handled and its temperature, as well as to wind conditions, and the frequency and duration of the loading task. The hazard presented to personnel depends upon the foregoing factors, as well as the toxic properties of the substance, the procedures/work practices followed, the protective equipment used, and the controls in place to reduce the likelihood of a release. Exposure to the materials handled can be reduced by metered loading, closed systems with off-gassing (gas or vapor) directed to an elevated vent or flare, vapor recovery systems, mechanical ventilation, and other means.

CONSIDERATIONS

Personnel performing product loading should be trained in the appropriate procedures/work practices required for the loading task, be aware of the hazardous properties of the materials being handled, and the appropriate measures to take for minimizing exposure during routine loading operations, as well as if a release occurs (e.g., spill). Protective equipment should be available to personnel and they should be trained in its use. A deluge shower/eyewash fountain should be readily accessible from the loading position. Signs should be posted indicating the appropriate controls for minimizing exposure and the emergency measures to take if there is contact with the material, or if a spill occurs.

Where practical, vapor/gas recovery systems could be provided to minimize a release during product loading. Personnel should be instructed to stand upwind of the loading position, extend the fill-pipe to the bottom of the tank (or as near as possible to it) to prevent splash loading, and to avoid standing directly above an open hatch (whether at metered loading, visual loading, or while topping-off). When catching samples during loading of ships/barges, the tank hatches and ullage covers should be kept closed, as much as practical.

Samples obtained during product loading should be properly identified and labels applied to the container so that health hazard information can be obtained, as necessary. Material safety data information (e.g., MSDSs) should be readily available. The samples should not be taken into an office or vehicle while awaiting transfer, or while being transferred to the laboratory.

If leakage is detected at a loading manifold, loading arm, hose connection, or other location, it should be reported promptly. Maintenance personnel should be aware of the hazards of the leaking material and provided appropriate protective equipment to use while effecting repairs.

If the material being handled contains hydrogen sulfide, special procedures are necessary to reduce the potential for exposure to this material, as well as to the stream vapor/gas. These measures may include assuring that personnel are aware of the potential for exposure to H_2S, disallowing loading personnel to work from a position at which exposure is likely, installing continuous monitors with alarm capability and, when appropriate, providing respiratory protective equipment (e.g., self-contained breathing unit), etc.

REFERENCES

Halder, C. A., et al, "Gasoline Vapor Exposures, Part 1, Characterization of Workplace Exposures", American Industrial Hygiene Association Journal, March 1986, pp. 164-177.

WORKPLACE HEALTH PROTECTION PROGRAM ELEMENT CHECKLIST ON BULK PRODUCT LOADING

		Yes	No	Don't Know	N/A
1.	Are loading instructions posted or readily available at product loading facilities?	___	___	___	___
2.	Are signs posted at loading facilities to alert personnel of the potential hazards associated with the materials being handled?	___	___	___	___
3.	Have loading personnel been made aware of the exposure hazards associated with materials being handled?	___	___	___	___
4.	Are vehicle operators aware of the health hazards associated with the materials being loaded/transported?	___	___	___	___
5.	Have all personnel doing loading been trained in the proper procedures/work practices to follow at product loading operations?	___	___	___	___
6.	Is appropriate protective equipment available to personnel who do loading:				
	. Hand protection (gloves)?	___	___	___	___
	. Face protection?	___	___	___	___
	. Skin protection?	___	___	___	___
	. Eye protection?	___	___	___	___
	. Respiratory protection?	___	___	___	___
7.	Have personnel been trained in the proper use of the available protective equipment?	___	___	___	___
8.	Do personnel stand upwind, as much as practical, during loading operations?	___	___	___	___
9.	Is metered loading in place at:				
	. Truck loading racks?	___	___	___	___
	. Railcar loading racks?	___	___	___	___
	. Marine wharves/docks?	___	___	___	___

N/A - Not Applicable

WORKPLACE HEALTH PROTECTION PROGRAM ELEMENT CHECKLIST ON BULK PRODUCT LOADING

	Question	Yes	No	Don't Know	N/A
10.	Are controls effective for minimizing worker exposure to vapors/gases at:				
	. Truck loading racks?	___	___	___	___
	. Railcar loading racks?	___	___	___	___
	. Marine wharves/docks?	___	___	___	___
11.	Are equipment leaks promptly repaired at loading operations?	___	___	___	___
12.	Is a pan positioned beneath each product manifold to contain spills?	___	___	___	___
13.	Does the person filling the transport take measures to minimize spillage of material during product loading?	___	___	___	___
14.	Are materials readily available at loading facilities to contain and clean up spills?	___	___	___	___
15.	Are samples of materials obtained during product loading operations not permitted to be taken into an enclosed area (field/rack enclosure, office, truck cab) before, or while being taken to the laboratory?	___	___	___	___
16.	Are samples appropriately identified to enable prompt reference to safety data information (e.g., an MSDS)?	___	___	___	___
17.	Are facilities provided for the collection and/or recycling of sample residues?	___	___	___	___
18.	Is the potential for contact with liquid product minimized during loading?	___	___	___	___
19.	Is illumination at loading areas (rack, manifold, etc.) adequate?	___	___	___	___
20.	Are facilities readily available to loading personnel to enable them to practice good personal hygiene?	___	___	___	___

N/A - Not Applicable

WORKPLACE HEALTH PROTECTION PROGRAM ELEMENT CHECKLIST ON BULK PRODUCT LOADING

		Yes	No	Don't Know	N/A
21.	Do loading personnel adhere to good personal hygiene practices?	___	___	___	___
22.	Are product loading locations provided with a deluge shower/eyewash fountain, as warranted?	___	___	___	___
23.	Are emergency switches readily available to appropriate personnel (dispatch office, loading point) to shut down product transfers if an emergency arises?	___	___	___	___
24.	Are product loading lines inspected periodically to identify defects and minimize leakage?	___	___	___	___

N/A - Not Applicable

WORKPLACE HEALTH PROTECTION PROGRAM ELEMENT
ON
CONFINED SPACE ENTRY

WORKPLACE HEALTH PROTECTION PROGRAM ELEMENT ON CONFINED SPACE ENTRY

OBJECTIVE

To provide information for reducing the potential for adverse health effects to occur among personnel who must enter and work in confined spaces.

BACKGROUND

A confined space is one which has limited openings for entry and exit, is not intended for continuous occupancy, and has inadequate natural ventilation. A hazardous atmosphere may exist in the space or develop over time. Examples of confined spaces include process vessels, utility vaults, storage tanks, pump rooms on tankers, boilers, and others which meet the above criteria.

Adverse health effects may occur when personnel enter into and/or work in some confined spaces. Such effects may result from an oxygen deficiency, or the presence of a hazardous material or a physical hazard in the confined space. Incidents may occur due to conditions which exist when the entry is made, or as a result of the build-up of contaminant concentration, or a reduction in oxygen level during work activities. Accidents are often due to a lack of training of personnel involved in confined space work; failure to recognize the potential hazards associated with the work; inadequate ventilation; poor procedures or work practices; not locking-out a source of ionizing radiation; inadequate testing of the atmosphere in the confined space; misinterpretation of monitoring results; or from other causes.

It is possible to "enter" a confined space without going inside. Serious accidents have occurred when an individual peered into an enclosure which contained a low oxygen concentration or a high level of a toxic gas (e.g., H_2S, CO, etc.). In such situations, the individual broke the plane of the space and was exposed to the hazardous condition in doing so. Complete entry was not necessary for the adverse effect to occur.

Unacceptable environmental conditions in a confined space may be the result of sources of contamination within, as well as from outside the space. Vapors can be released from scale or sludge while personnel are working inside the space. Mechanical ventilation for providing air to the space may draw in airborne contaminants if there are sources nearby. This must be avoided by proper placement of the fan and elimination of sources of contamination. Operations carried out in the space, such as welding, cleaning, metal inspection, and others, can result in contaminant generation that may lead to the development of an inhalation hazard. Inert gas, (i.e., lacking oxygen) may be purposely put into a confined space, such as a tank, to displace vapors and/or oxygen, thereby presenting a potential hazard to those who enter the space, as well as those outside near open manways. Heat build-up, due to solar heating of an enclosed space, may give rise to an excessive heat stress situation within the space if entry is made. Use of air-operated power tools supplied with process or utility air of non-respirable quality, or with an inert gas such as nitrogen, can create a health hazard in a confined space atmosphere.

CONSIDERATIONS

Permit systems should address both the health and safety concerns associated with confined space work. Including these considerations in the permit issuance system is an effective tool for reducing the exposure potential and health risks associated with confined space work. A specific permit should be required for entry by contractors or Mobil personnel into, and work in, a confined space.

A written confined space entry program, including a permit procedure, should be developed. Personnel involved in such entries should be trained in proper work practices, hazard awareness, use of protective equipment, and permit procedures. Personnel responsible for signing entry permits must be qualified and identified.

Equipment used to test the confined space environment must be in proper calibration, tested before use, and operated properly by personnel trained in its use, as well as in the interpretation of results. When necessary, appropriate protective equipment must be provided personnel entering the space and they must be trained in its use. An individual should be stationed outside a confined space to serve as a back-up when work is being done inside. If an emergency arises help should be summoned to assist personnel in the confined space. An emergency evacuation and rescue procedure should be a part of the written confined space entry program.

Employee and contractor personnel who enter confined spaces must be aware of all actual and potential hazards associated with each entry. The level of oxygen, flammables, and toxic substances that are present in the space should be determined before the permit is issued and entry permitted. Continuous monitoring of a confined space environment is often advisable while work is being carried out. This may involve the continuous measurement of the oxygen level, or the concentration of some toxic substances (H_2S, CO, benzene, etc.). The individual in charge of the entry should be identified, as well as the duties of each individual involved in the entry. The period for which the permit is to be valid must be identified on the permit, along with specific precautions, protective equipment requirements, monitoring results, approval signatures, need for follow-up or continuous monitoring, as well as other relevant information.

REFERENCES

"Guidelines for Work in Confined Spaces in the Petroleum Industry", Publication 2217 (Washington, D.C.: American Petroleum Institute, 1984).

"Guideline for Work in Confined Spaces in the Petroleum Industry", Publication 2217A (Washington, D.C.: American Petroleum Institute, 1987).

Rekus, J.F., "Invisible Confined Space Hazards Require Comprehensive Entry Program", Occupational Health & Safety Journal, August 1990, pp. 38-51.

"Guide for Inspection of Refinery Equipment", Chapter V, Preparation of Equipment for Safe Entry and Work (Washington, D.C.: American Petroleum Institute, December 1978).

Garrison, R. P., R. Nabar and M. Erig, "Ventilation to Eliminate Oxygen Deficiency in a Confined Space", Applied Industrial Hygiene Journal, January 1989, pp. 1-11.

Garrison, R. P., R. Nabar and M. Erig, "Ventilation to Eliminate Oxygen Deficiency in a Confined Space", Part II, Applied Industrial Hygiene Journal, October 1989, pp. 260-268.

Garrison, R. P. and M. Erig, "Ventilation to Eliminate Oxygen Deficiency in a Confined Space", Part III, Applied Industrial Hygiene Journal, February 1991, pp. 131-140.

"Safety Requirements for Confined Spaces", ANSI Z117.1-1989 (New York: American National Standards Institute, 1989).

WORKPLACE HEALTH PROTECTION PROGRAM ELEMENT CHECKLIST ON CONFINED SPACE ENTRY

		Yes	No	Don't Know	N/A
1.	Is there a written confined space entry program for the facility?	___	___	___	___
2.	Are confined spaces that are to be entered identified as such if not readily apparent?	___	___	___	___
3.	Is there a permit system in effect for entering a confined space?	___	___	___	___
4.	Is there a written procedure for completing the confined space entry permit?	___	___	___	___
5.	Are responsibilities for preparing a confined space for entry clearly defined?	___	___	___	___
6.	Does the entry permit procedure designate:				
	. Who is to issue the permit?	___	___	___	___
	. Who will test the space?	___	___	___	___
	. Who will approve the entry?	___	___	___	___
	. What contaminant concentrations are acceptable?	___	___	___	___
	. What personal protective equipment is to be used?	___	___	___	___
	. Time period the permit is valid?	___	___	___	___
7.	Are personnel who work in confined spaces trained to recognize the potential health hazards associated with the work?	___	___	___	___
8.	Are contractors who do confined space work aware of the potential hazards and the relevant facility procedures/work practices?	___	___	___	___

N/A - Not Applicable

WORKPLACE HEALTH PROTECTION PROGRAM ELEMENT CHECKLIST ON CONFINED SPACE ENTRY

		Yes	No	Don't Know	N/A
9.	Is oversight carried out to ensure that employees/contractors adhere to permit conditions and accepted procedures/work practices associated with confined space work?	___	___	___	___
10.	Is training provided personnel who:				
	. Authorize confined space entries?	___	___	___	___
	. Test the environment for entry?	___	___	___	___
	. Enter the space to do work?	___	___	___	___
	. Provide back-up support (attendant)?	___	___	___	___
	. Respond if an emergency arises?	___	___	___	___
11.	Is the equipment which is available for testing confined space environments:				
	. Appropriate for the substances encountered?	___	___	___	___
	. Adequately maintained?	___	___	___	___
	. Periodically calibrated?	___	___	___	___
	. Calibration records maintained?	___	___	___	___
	. Properly used?	___	___	___	___
12.	When appropriate, is testing of the confined space environment carried out with the ventilation system off? (NOTE - Sometimes exhaust air is tested to determine contaminant concentration.)	___	___	___	___
13.	Are personnel who test confined space environments knowledgeable in the interpretation of monitoring results?	___	___	___	___
14.	Where required, are confined spaces effectively isolated before a permit is issued for entry?	___	___	___	___

N/A - Not Applicable

WORKPLACE HEALTH PROTECTION PROGRAM ELEMENT CHECKLIST ON CONFINED SPACE ENTRY

		Yes	No	Don't Know	N/A
15.	Are internal structures that may impede exit or emergency rescue recognized and either eliminated, or otherwise provided for before entry is permitted?	___	___	___	___
16.	Are acceptable exposure periods established for work in a hot, confined environment?	___	___	___	___
17.	Has an upper temperature limit been established for entry into a furnace (firebox) or other hot environment for:				
	. Inspection?	___	___	___	___
	. Maintenance work?	___	___	___	___
	. Repair work?	___	___	___	___
18.	Is effective eye protection provided to personnel to prevent contact with acidic particulates/ dust during work inside furnaces/boilers, etc.?	___	___	___	___
19.	Is monitoring of the air in a confined space continued while work is proceeding?	___	___	___	___
20.	Is appropriate protective equipment available, when necessary, and used by personnel involved in confined space work:				
	. Respiratory protection?	___	___	___	___
	. Eye protection?	___	___	___	___
	. Skin/face/hand/foot protection?	___	___	___	___
21.	Is there proper illumination in the confined space for the work being performed?	___	___	___	___
22.	Is ventilation provided for the confined space:				
	. General mechanical ventilation:				
	Supply?	___	___	___	___
	Exhaust?	___	___	___	___

N/A - Not Applicable

WORKPLACE HEALTH PROTECTION PROGRAM ELEMENT CHECKLIST ON CONFINED SPACE ENTRY

	Yes	No	Don't Know	N/A
. Local mechanical exhaust for application at the contaminant source?	___	___	___	___
. Natural ventilation?	___	___	___	___
23. Is the ventilation that is provided adequate for the confined space work?	___	___	___	___
24. Is the potential for re-entry of exhausted contaminants always assessed when confined space work is done?	___	___	___	___
25. Are ionizing radiation sources on equipment to be entered locked out before entry is permitted?	___	___	___	___
26. Is a back-up person always present when a confined space entry is made?	___	___	___	___
27. Does the back-up person maintain communication with person(s) in the confined space?	___	___	___	___
28. Does back-up person have means available to summon assistance?	___	___	___	___
29. Are trained personnel available (either on or off-site) to respond promptly to confined space emergencies?	___	___	___	___
30. Are in-force entry permits checked to ensure they:				
. Are posted at the work location?	___	___	___	___
. Are properly signed, dated, authorized, etc.?	___	___	___	___
. Properly identify the individual in charge of entry?	___	___	___	___
. Are properly filled out?	___	___	___	___
. Are being adhered to?	___	___	___	___

N/A - Not Applicable

WORKPLACE HEALTH PROTECTION PROGRAM ELEMENT CHECKLIST ON CONFINED SPACE ENTRY

		Yes	No	Don't Know	N/A
31.	Are confined space entry re-testing requirements adhered to when permits are issued for longer than the normal permitted period (e.g., 4 hours, 8 hours)?	___	___	___	___
32.	Is there a capability to carry out continuous monitoring within the confined space for:				
	. Oxygen?	___	___	___	___
	. Combustibles/flammables?	___	___	___	___
	. Toxics?	___	___	___	___
33.	Is entry onto the roof of an external floating roof tank considered a confined space entry?	___	___	___	___
34.	Does entry onto the roof of an internal floating roof tank require special testing considerations and a permit?	___	___	___	___
35.	Are there written procedures for entry onto the roof of internal floating roof tanks?	___	___	___	___

N/A - Not Applicable

WORKPLACE HEALTH PROTECTION PROGRAM ELEMENT
ON
FUGITIVE EMISSIONS

WORKPLACE HEALTH PROTECTION PROGRAM ELEMENT ON FUGITIVE EMISSIONS

OBJECTIVE

To identify the need to quantify releases from fugitive hydrocarbon emission sources in petroleum and petrochemical operations, provide information on a procedure to accomplish this, and measures to take to minimize fugitive emissions and their contribution to personnel exposures.

BACKGROUND

The objective of workplace health protection programs is to anticipate, recognize, evaluate, and control those factors in the work environment which can cause adverse health effects, including occupational diseases, chronic or irreversible tissue damage, irritation or narcosis of sufficient degree to result in accidental injury, impairment of self rescue, or reduced efficiency. During the evaluation of exposures to hazardous agents (chemical or biological) it is of interest to identify and address those sources of contamination which contribute to the exposure and, if possible, quantify the contribution of each, so that appropriate recommendations can be made to reduce exposures, when necessary.

In the course of evaluating employee exposure to hydrocarbon vapors in a petroleum refinery or petrochemical facility, it is difficult, after the fact, to identify the source(s) of contamination that contributed to the exposure. Potentially, there could be a large number. One approach that may be used to identify these in process operations is to follow the employee during rounds, while catching samples, pulling blinds, opening manways, draining equipment, and performing other tasks. Such activities often account for a significant portion of the hydrocarbon vapor exposure that an individual experiences. There are, however, other identifiable sources of airborne contamination that contribute to the exposure. Some may be fugitive emission sources. These include vents, pump seals, pressure regulators, valves, flanges, sample points, and others. Releases here are often the result of vibration, improper closure of equipment, inadequate pump packing, worn gaskets, and other maintenance related causes.

It is of interest to quantify the magnitude of releases from fugitive emission sources during periods when personal sampling is being conducted on facility personnel. It is most appropriate when a comprehensive survey is being conducted. If exposures are considered high, recommendations can be presented to reduce the contribution from significant fugitive emission sources identified in the monitoring effort.

Not all fugitive emission sources need to be evaluated. Emissions from those which handle the contaminant of concern, and those which experience has shown to be significant contributors, should be evaluated.

CONSIDERATIONS

A method for determining the hydrocarbon vapor/gas contribution from fugitive emission sources to the general contamination of an area is to use a portable, direct reading, sample draw type hydrocarbon vapor/gas monitor. Measurements are made at about 1 inch from pump or compressor seals, valve seals, flanges, safety vent pipes, surface drains, and other sources. The object is to identify significant sources of release. Hydrocarbon vapor/gas concentrations above 500 parts per million at the indicated position are generally considered significant. The fugitive emission level of concern, however, will depend on the toxicity of the substance, its exposure limit, and the likelihood for a person to be exposed to it.

WORKPLACE HEALTH PROTECTION PROGRAM ELEMENT CHECKLIST ON FUGITIVE EMISSIONS

		Yes	No	Don't Know	N/A
1.	Have fugitive emission sources been evaluated relative to their contribution to employee hydrocarbon vapor/gas exposures?	___	___	___	___
2.	Are continuous hydrocarbon monitors in operation to detect significant releases of flammable gases or vapors that can cause adverse health effects?	___	___	___	___
3.	Are the continuous monitors adequately maintained to ensure effective operation?	___	___	___	___
4.	Are there sufficient continuous monitors in operation to identify releases that may present an exposure hazard to the health of personnel?	___	___	___	___
5.	Is an investigation conducted to identify sources of release when a continuous monitor alarm occurs?	___	___	___	___
6.	Is piping adequately supported to prevent vibrations, loosening of flange connections and consequent fugitive emissions?	___	___	___	___
7.	Are flanges connected properly (all bolts the proper type, all in place, and tight)?	___	___	___	___
8.	Is there an absence of ice (often evidence of leakage) on pumps handling gases and high vapor pressure liquids?	___	___	___	___
9.	Are pump seal leaks promptly eliminated when identified?	___	___	___	___
10.	Is piping to and from vessels and sample points adequately supported to prevent leaks at connections?	___	___	___	___

N/A - Not Applicable

WORKPLACE HEALTH PROTECTION PROGRAM ELEMENT CHECKLIST ON FUGITIVE EMISSIONS

		Yes	No	Don't Know	N/A
11.	Are there closed drain systems at sample locations for disposal of material from line flushing/container rinsing?	___	___	___	___
12.	Are sample container rinsings properly disposed?	___	___	___	___
13.	Are valves at sample points properly closed to prevent leakage?	___	___	___	___
14.	Are vents from process analyzers sufficiently high to minimize contamination at grade or other potentially occupied location?	___	___	___	___
15.	Are compressor seal oil drums vented at sufficient height to prevent contamination at grade or other potentially occupied location?	___	___	___	___
16.	Are double seals specified for equipment where single seals have been inadequate to prevent releases?	___	___	___	___
17.	Are safety valve releases vented such that the potential for contamination at grade or other occupied location is minimized?	___	___	___	___
18.	Are drains at grade flushed with sufficient water to carry away liquid hydrocarbons and thereby reduce emissions?	___	___	___	___
19.	Are emissions adequately controlled so that the use of steam is not needed to dissipate vapors/gases that are released from equipment?	___	___	___	___

N/A - Not Applicable

WORKPLACE HEALTH PROTECTION PROGRAM ELEMENT CHECKLIST ON FUGITIVE EMISSIONS

		Yes	No	Don't Know	N/A
20.	Are pump pads provided with adequate surface drains to facilitate removal of spillage/leakage?	___	___	___	___
21.	Are surface areas adequately sloped to provide for prompt drainage of liquids to a drain or sump?	___	___	___	___
22.	Are used filters and residue from basket strainers effectively disposed so that vapor releases or surface contamination from them are not a source of fugitive emissions?	___	___	___	___
23.	Are pump, tank, drum, etc. drains piped through a closed system to a catch basin or sump?	___	___	___	___
24.	Are manholes and sumps provided with covers and vented to a satisfactory location?	___	___	___	___

N/A - Not Applicable

WORKPLACE HEALTH PROTECTION PROGRAM ELEMENT
ON
HAZARD WARNING SIGNS

WORKPLACE HEALTH PROTECTION PROGRAM ELEMENT ON HAZARD WARNING SIGNS

OBJECTIVE

To provide information for ensuring the effectiveness of health hazard warning signs.

BACKGROUND

Posting of signs is a method to alert personnel of the existence of a present or potential health hazard in an area. Signs are most effective when they are attention getting, properly located, readily visible, and worded for easy understanding. The message presented must be clear and unequivocal. Where appropriate, signs should be properly lighted for visibility at night.

The words used to alert personnel of a health hazard must be consistently applied. "Danger" signs are for use where a clear and present danger exists. "Caution" signs are used to warn against potential hazards or unsafe practices. "Instructional" signs are used when there is a need for general instructions and suggestions relative to a health concern.

Some organizations have established company-wide policies on signs. This enables the development of a program which can serve as a comprehensive guide for sign color, form, application, and usage throughout all facilities.

CONSIDERATIONS

Groups responsible for signage within each facility should ensure signs that are in place are adequate with respect to number, location, color, wording and condition. Where necessary, they should be replaced to conform to a standard form. There should be no variation in the design, color, or wording of signs posted to warn of a specific hazard in a facility. That is, all signs should be the same color and employ the same wording for the same hazard.

The location of signs should be visible to potentially affected personnel, and be placed ahead of a hazard to provide individuals coming into an area adequate time to respond to the warning. Signs should not be placed on doors, windows or other moveable surface. Sign wording should make a positive rather than a negative statement and they should be consistent for a specific hazard throughout the facility. An adequate number of signs should be posted at accessways to a hazardous area in order to alert personnel to the hazard that may be encountered and of the action to be taken to reduce risk.

Where necessary, local regulations should be complied with regarding sign coloring, wording, etc.

REFERENCE

"USA Standard Specification For Accident Prevention Signs," USAS Z35.1-1968 (New York: American National Standards Institute, 1968).

WORKPLACE HEALTH PROTECTION PROGRAM ELEMENT CHECKLIST ON HAZARD WARNING SIGNS

	Yes	No	Don't Know	N/A
1. Is the color, size, and wording of signs consistent with company guidelines and/or applicable regulations?	___	___	___	___
2. Are signs for alerting personnel to a potential or present hazard:				
. Posted at appropriate locations?	___	___	___	___
. Placed to provide personnel time to respond to the warning?	___	___	___	___
. Provided with a clearly stated message?	___	___	___	___
. Clean and legible?	___	___	___	___
3. Is the wording and coloring of warning signs for a specific hazard the same throughout the facility:				
. Emergency self-contained breathing equipment?	___	___	___	___
. Deluge showers/eyewash fountains?	___	___	___	___
. Hydrogen sulfide containing gas streams?	___	___	___	___
. Continuous monitor alarms?	___	___	___	___
. Breathing air connections?	___	___	___	___
. Personal protective equipment for routine use?	___	___	___	___
. Access to tanks containing sour materials?	___	___	___	___
. Leaded gasoline storage tanks?	___	___	___	___
. Ionizing radiation sources?	___	___	___	___
. Eye protection use required?	___	___	___	___
. Hearing protection use required?	___	___	___	___
. Others? (______________________)	___	___	___	___

N/A - Not Applicable

WORKPLACE HEALTH PROTECTION PROGRAM ELEMENT
ON
LABORATORY OPERATIONS

WORKPLACE HEALTH PROTECTION PROGRAM ELEMENT ON LABORATORY OPERATIONS

OBJECTIVE

To provide information for evaluating procedures and work practices, assessing the performance of engineering controls (e.g., general and local exhaust ventilation), and identifying personal protective equipment needs to minimize employee exposure to the substances handled in laboratory operations.

BACKGROUND

Physical and chemical analyses are routinely carried out in laboratories to support petroleum and petrochemical manufacturing, marketing, and other operations. Such tests are performed to ensure that raw materials, intermediates, and products meet specifications. The scope of analytical procedures carried out, the number of employees engaged in this work, the chemicals used and the frequency and duration of exposures to them depends upon the nature of the operations, facility size, as well as the type and number of samples submitted to the laboratory per day.

The composition of raw materials, intermediates, and final products vary. They may contain chemical constituents that can pose potential health hazards to personnel. Chemicals that exist in significant amounts may include benzene, 1,3-butadiene, n-hexane, hydrogen sulfide, polynuclear aromatic compounds and others. In addition, chemicals used in analytical procedures have the potential to cause adverse health effects as a result of skin contact or inhalation of the mist, vapor, gas, or fume emitted during handling and analysis. Toxic vapors, gases, or mists may be evolved in some analyses and these must be exhausted to prevent exposure to them. Corrosive materials used in the laboratory can damage the skin and eyes if contact occurs. Thus, a variety of controls, including engineering and administrative methods, and use of personal protective equipment are needed to effectively reduce exposure potential.

CONSIDERATIONS

All substances (reagents, materials to be analyzed, gases used in analytical equipment, process samples, etc.) that are brought into the laboratory should be properly labeled and a material safety data sheet (MSDS) be available for each. Purchased chemicals should have label statements indicating the potential hazards associated with handling and use, and date of receipt. If appropriate, they should be dated when opened and the expiration date noted. Chemicals should be stored so that the likelihood for incompatibilities to occur with other materials is minimized.

Chemical stockrooms should be maintained at the proper temperature and humidity. Laboratory operations should not be conducted in a storage area. Storage should be arranged in accordance with the chemicals' reactive properties, not alphabetically. The stockroom should be provided with adequate ventilation so that it is at slight positive pressure with respect to surroundings. When warranted, a deluge shower/eyewash fountain should be provided here.

Personnel working in a laboratory should receive training regarding the potential health hazards associated with materials used, and of the proper procedures for the handling and storage of them.

Proper facilities should be available for receiving and storing samples. A sample receipt room, provided with a mechanical exhaust ventilated hood, should be available. Sample retain rooms and chemical storage areas should be provided with adequate mechanical ventilation to prevent a build-up of airborne contaminants. Flammable liquids should be stored, as much as practical, in appropriately labeled flammable liquid storage cabinets.

Eating, drinking, and smoking are not to be permitted in the laboratory. A separate designated area should be provided. In addition, adequate personal hygiene facilities should be available.

Personal protective equipment should be on hand and available for employee use. Personnel who may need to use the protective equipment should be trained in its use.

Eye protection should be worn by all personnel entering the laboratory. Signs indicating this requirement should be posted at entrances. Contact lens use by personnel involved in laboratory operations should be reviewed and a policy established regarding the acceptability of their use here.

Deluge shower/eyewash fountain facilities should be readily available for emergency use. They should be tested periodically and personnel trained in the proper use of this equipment.

Laboratory personnel who may use, or be required to use, respiratory protection should be included in the respirator program (be medically approved to use such equipment, trained in its use, and fit tested). Appropriate equipment should be readily available to personnel.

Analytical procedures and handling of chemicals should be carried out, if at all possible, inside exhaust ventilated hoods. Ventilation should be adequate to contain and remove the contaminants released or generated in the work being performed. An average face velocity of 100 feet per minute is typically recommended for general laboratory hoods. Lower face velocities may be acceptable if there is good distribution of make-up air, low traffic past the hood, and the materials being handled are of low toxicity. Hood ventilation should be evaluated periodically (e.g., annually) to assure it is adequate. Exhaust stacks should be the concentric design or other non-downward deflecting type. Conventional rain caps should not be used.

General laboratories located in buildings which also serve other purposes should be at negative pressure with respect to surroundings so that airflow is into the laboratory and not out (e.g., into hallways, offices, etc). For some laboratories, such as those in which biohazards may exist, room tightness and air-supply/exhaust balance is essential. In all cases, provisions should be made to supply adequate tempered make-up air to the laboratory.

Where fixed Halon or carbon dioxide (CO_2) fire extinguishing systems are in place, personnel should be aware of the potential hazards that may occur if these are discharged. High pressure CO_2 and Halon releases are accompanied by loud noise, as well as the potential to cause materials to fall off shelves, laboratory benches, etc. Carbon dioxide concentrations necessary for fire extinguishment result in depletion of the oxygen level in the space into which the CO_2 gas is released. The oxygen concentration can be well below that necessary to sustain life. Halon systems are designed to achieve a concentration of approximately 7 percent. With Halon at this concentration, the resulting oxygen level (above 19%) would be sufficient to support life. In a fire situation, Halon can be decomposed to form compounds more toxic than the parent substance. Thus, it would be advisable to test the air before entry is permitted following a CO_2 or Halon system discharge. Safe procedures should be written for responding to CO_2 or Halon releases and personnel made aware of the potential health hazards when responding to an emergency situation where exposure to these contaminants may occur.

Adequate illumination should be provided on laboratory benches, in hoods, offices, and along aisleways in the laboratory so that work can be carried out safely and without creating eye strain. Emergency lighting should be provided throughout the laboratory and it should be tested periodically to ensure operability and adequacy.

Vacuum pump exhaust should be discharged through a vent pipe or tube to an acceptable (above roof) outside location or into a continuously operated exhaust hood. If a vacuum is pulled on a system containing mercury, the vacuum pump oil can become contaminated with mercury. Appropriate procedures should be established for removal and disposal of this oil since it may contain mercury.

If an X-ray diffraction unit is operated in the laboratory, or other sources of ionizing radiation are used here, a radiation exposure control program should be established. Radiation sources may require a license and/or registration with regulatory agencies. Each source/device should be properly labeled. Periodic wipe tests should be carried out to ensure sealed source integrity is maintained. Records of wipe test results, registrations, and other documentation related to each source must be properly maintained. Where appropriate, personnel working with sources of ionizing radiation should be monitored with film badges or thermoluminescent dosimeters (TLDs) to determine their exposure to ionizing radiation. Exposure results should be maintained. Depleted radiation sources should be properly disposed.

Compressed gas cylinders used to supply gases to the laboratory should be properly handled and stored. Cylinder storage and use inside the laboratory is to be discouraged. Cylinder size should be selected based on anticipated need, availability, safe use considerations, consequence of total release, etc. Personnel involved in connecting and disconnecting cylinders to regulators and gas transfer lines should be properly trained. Gas transfer lines and pressure fittings should be an appropriate type for preventing leakage or accidental releases. Leak tests should be conducted to ensure the integrity of gas transfer systems.

In some facilities, laboratory employees collect process samples from operating equipment. These personnel should be aware of the potential hazards associated with entry into the units/areas/locations where samples are to be obtained, and of the hazards associated with the materials being sampled. Acceptable sampling and handling procedures should be developed and followed. Collected samples should be transported in the proper containment, be adequately identified, and stored so that the potential for breakage/leakage is minimized. If leakage does occur, the released material should be controlled so that it does not present a hazard to employees, the facility, or the environment. Acceptable procedures should be established and followed in disposing of samples and retains.

A written spill control procedure and emergency response plan should be in place for responding to laboratory emergencies. The plan should be posted and known by all laboratory personnel. Responsibilities of personnel should be clearly spelled out and periodic emergency drills conducted (evacuation, rescue, medical emergency, etc.).

A laboratory safety committee should be established, where warranted, with responsibility for developing safe procedures/work practices; assuring established guidelines are adhered to; investigating accidents and near misses; and carrying out periodic inspections of the facility. In the absence of such a committee, personnel should be designated to be responsible for developing procedures/work practices to follow in laboratory operations to ensure workplace health hazards are effectively controlled.

REFERENCES

Dux, J. P. and R. F. Stalzer, "Managing Safety in the Chemical Laboratory" (New York: Van Nostrand Reinhold Company, 1988).

"Safety in the Chemical Laboratory" (Easton, PA: American Chemical Society, 1974).

Di Berardinis, L., et al, "Guidelines for Laboratory Design: Health and Safety Considerations" (New York: John Wiley & Sons, 1987).

Young, J. A., Ed., "Improving Safety in the Chemical Laboratory: A Practical Guide" (New York: John Wiley & Sons, 1987).

"Biohazards Reference Manual" (Akron, OH: American Industrial Hygiene Association, 1985).

"Handbook of Compressed Gases", 2nd ed. (New York: Van Nostrand Reinhold Company, 1980).

Furr, A. K., Ed., "Guide to Safe Practices in Chemical Laboratories" (Boca Raton, FL: CRC Press, Inc., 1989).

"Industrial Ventilation - A Manual of Recommended Practice", Latest Edition (Lansing, MI: American Conference of Governmental Industrial Hygienists).

Abrams, D. S., et al, "An Evaluation of the Effectiveness of a Recirculating Laboratory Hood", American Industrial Hygiene Association Journal (January 1986), pp. 22-26.

Bradford, W. J., "Storage and Handling of Chemicals", National Fire Protection Handbook, 16th Ed, Section 12 (Quincy, MA: National Fire Protection Association, 1986), pp. 6-11.

Finley, K. J., "The Business of Labs", Asbestos Issues (August 1990), pp. 22-26.

Beltramini, H. P., "Protection for Laboratories", National Fire Protection Handbook, 16th Ed., Section 11 (Quincy, MA: National Fire Protection Association, 1986), pp. 49-54.

Clarke, A. N., et al, "The First Step Toward Total Quality Assurance", American Environmental Laboratory (Oct. 1990), pp. 9-14.

"Pocket Guide to Chemical Hazards", DHHS (NOISH) Publication No. 90-117, (Cincinnati, OH: National Institute for Occupational Safety and Health, 1990).

"Method of Testing Performance of Laboratory Fume Hoods", ANSI/ASHRAE 110-1985 (New York: American National Standards Institute, 1985).

WORKPLACE HEALTH PROTECTION PROGRAM ELEMENT CHECKLIST ON LABORATORY OPERATIONS

		Yes	No	Don't Know	N/A
1.	Is there a laboratory safety committee in place?	___	___	___	___
2.	Is the charter of the laboratory safety committee clearly defined?	___	___	___	___
3.	Is there a written workplace health protection/safety manual for the laboratory?	___	___	___	___
4.	Have laboratory personnel been provided a copy of the laboratory workplace health protection/safety manual?	___	___	___	___
5.	Are laboratory procedures (test methods), design changes, etc., relative to workplace health protection issues, reviewed and approved by various designated functions (e.g., industrial hygiene, safety, etc.)?	___	___	___	___
6.	Are periodic laboratory inspections carried out to identify unsatisfactory conditions?	___	___	___	___
7.	Are the workplace health protection program deficiencies noted in operation reviews documented?	___	___	___	___
8.	Is there a follow-up procedure to ensure that workplace health related deficiencies noted in inspections/surveys have been corrected?	___	___	___	___
9.	Is there a purchase review procedure for new chemicals being requisitioned for use in the laboratory?	___	___	___	___

N/A - Not Applicable

WORKPLACE HEALTH PROTECTION PROGRAM ELEMENT CHECKLIST ON LABORATORY OPERATIONS

		Yes	No	Don't Know	N/A
10.	Does the laboratory safety committee or responsible individual ensure that workplace health exposure controls are adequate for new chemicals to be used?	___	___	___	___
11.	Is there a review procedure to determine whether:				
	. It is acceptable to work alone in the lab?	___	___	___	___
	. It is acceptable to carry out a specific laboratory procedure when working alone?	___	___	___	___
12.	Are all chemicals received by the laboratory:				
	. Properly labeled?	___	___	___	___
	. Dated when received?	___	___	___	___
	. Accompanied by appropriate hazard information (e.g., MSDS)?	___	___	___	___
	. Properly stored to prevent incompatibilities?	___	___	___	___
13.	If chemicals received by the laboratory are not properly labeled are there written procedures for personnel to follow to correct such deficiency?	___	___	___	___
14.	Are copies of the analytical procedures that are carried out available at the laboratory?	___	___	___	___
15.	Are all personnel who work in the laboratory provided training advising them of:				
	. Hazards associated with chemicals handled?	___	___	___	___
	. The proper way to handle and store chemicals?	___	___	___	___
	. Use of available protective equipment?	___	___	___	___

N/A - Not Applicable

WORKPLACE HEALTH PROTECTION PROGRAM ELEMENT CHECKLIST ON LABORATORY OPERATIONS

	Yes	No	Don't Know	N/A
• Emergency procedures?	___	___	___	___
• Responsibilities in emergency situations?	___	___	___	___
• Potential chemical incompatibilities?	___	___	___	___
16. Where applicable, are laboratory personnel adequately trained to obtain process samples?	___	___	___	___
17. Is hazard information, such as MSDSs, readily available to employees?	___	___	___	___
18. Is the proper protective equipment available to laboratory personnel:				
• Eye protection?	___	___	___	___
• Face protection (e.g., face shield)?	___	___	___	___
• Gloves (appropriate for materials handled)?	___	___	___	___
• Respiratory protection?	___	___	___	___
• Eyewash fountain?	___	___	___	___
• Emergency shower facility?	___	___	___	___
• Exhaust ventilated hood?	___	___	___	___
19. Is there a "safety drawer" in the laboratory, (i.e., contains equipment for emergency use)?	___	___	___	___
20. Are flammable liquids stored in a flammable liquid storage cabinet?	___	___	___	___
21. Is the flammable liquid storage cabinet provided with adequate exhaust ventilation?	___	___	___	___
22. Are laboratory analyses carried out inside laboratory hoods (as much as practical)?	___	___	___	___
23. Is there a review procedure for determining whether a laboratory procedure should be carried out in a hood?	___	___	___	___

N/A - Not Applicable

WORKPLACE HEALTH PROTECTION PROGRAM ELEMENT CHECKLIST ON LABORATORY OPERATIONS

		Yes	No	Don't Know	N/A
24.	Have laboratory personnel been instructed in the proper use of exhaust ventilated hoods?	___	___	___	___
25.	Are laboratory hoods equipped with airfoil edges?	___	___	___	___
26.	Are laboratory hoods properly located (e.g., not near a frequently used door, out of high traffic area, not directly below an air supply vent, etc.)?	___	___	___	___
27.	Are laboratory hoods provided with a spill lip?	___	___	___	___
28.	Is hood housekeeping acceptable?	___	___	___	___
29.	Is hood ventilation performance evaluated periodically?	___	___	___	___
30.	Is laboratory hood ventilation adequate for the work being carried out?	___	___	___	___
31.	Is each laboratory hood identified as to:				
	. Date of last ventilation test?	___	___	___	___
	. Result of last ventilation test?	___	___	___	___
	. Sash position when test conducted?	___	___	___	___
	. Sash position to achieve adequate face velocity?	___	___	___	___
32.	Are hoods equipped with a flow indicator?	___	___	___	___
33.	Are sinks which are used for sample disposal provided with exhaust ventilation?	___	___	___	___
34.	Are laboratory areas in multi-use buildings at negative pressure with respect to surroundings (corridors, office areas, etc.)?	___	___	___	___

N/A - Not Applicable

WORKPLACE HEALTH PROTECTION PROGRAM ELEMENT CHECKLIST ON LABORATORY OPERATIONS

		Yes	No	Don't Know	N/A
35.	Is the air supply to laboratory areas adequate:				
	. General laboratory?	___	___	___	___
	. Bottle wash room?	___	___	___	___
	. Sample retain room/facility?	___	___	___	___
	. Other locations(___________)?	___	___	___	___
36.	Are smoke tubes used to evaluate turbulence around laboratory hoods and identify the need to correct/minimize it?	___	___	___	___
37.	Is use of eye protection required for all personnel entering the laboratory?	___	___	___	___
38.	Has contact lens use by laboratory personnel been reviewed?	___	___	___	___
39.	Is contact lens use prohibited in the laboratory?	___	___	___	___
40.	Are deluge shower/eyewash fountain operational checks made periodically and results recorded?	___	___	___	___
41.	Are deluge showers/eyewash fountains provided with potable water?	___	___	___	___
42.	Are deluge showers/eyewash fountains provided with adequate drainage?	___	___	___	___
43.	Is an alarm activated if a deluge shower/eyewash fountain is used?	___	___	___	___
44.	Does the facility's respirator program, (where appropriate) include training of laboratory personnel?	___	___	___	___

N/A - Not Applicable

WORKPLACE HEALTH PROTECTION PROGRAM ELEMENT CHECKLIST ON LABORATORY OPERATIONS

	Yes	No	Don't Know	N/A
45. Is the respiratory protective equipment that is available to laboratory personnel properly maintained and readily available?	___	___	___	___
46. Is laboratory housekeeping adequate:				
. Benches?	___	___	___	___
. Aisles?	___	___	___	___
. Drawers?	___	___	___	___
. Hoods?	___	___	___	___
. Hood cabinets?	___	___	___	___
. Stockroom?	___	___	___	___
47. Are records related to laboratory radiation sources properly maintained:				
. Licenses?	___	___	___	___
. Registrations?	___	___	___	___
. Responsible person identified for each source?	___	___	___	___
. Leak test results?	___	___	___	___
. Source locations?	___	___	___	___
. Posting notifications?	___	___	___	___
. Disposal?	___	___	___	___
. Personal monitoring results?	___	___	___	___
48. Have the consequences of a power failure, or other emergency been considered in the hazard assessment of laboratory operations?	___	___	___	___
49. Is there a written emergency response plan for the laboratory?	___	___	___	___
50. Is there a written spill clean-up procedure applicable to laboratory operations?	___	___	___	___
51. Are spill kits readily available for use as needed?	___	___	___	___

N/A - Not Applicable

WORKPLACE HEALTH PROTECTION PROGRAM ELEMENT CHECKLIST ON LABORATORY OPERATIONS

	Yes	No	Don't Know	N/A
52. Are laboratory personnel who may be involved in spill/emergency response provided appropriate hazard awareness training for these activities?	___	___	___	___
53. Has a list of names been developed for emergency response in the laboratory and made available to personnel responsible for call out?	___	___	___	___
54. Do laboratory personnel involved in spill/emergency response know their responsibilities?	___	___	___	___
55. Is emergency response equipment that is available for laboratory incidents checked periodically to ensure its readiness for use?	___	___	___	___
56. Are continuous monitors operated in the laboratory, when appropriate, for detecting a toxic material release?	___	___	___	___
57. Is there an established basis for determining the need to install a continuous monitor in the laboratory (e.g., use of highly toxic material with poor warning properties, large volume used, high potential for release, etc.)?	___	___	___	___
58. Are continuous monitors periodically calibrated by a qualified person?	___	___	___	___
59. Are concentric-type or other non-downward deflecting rain caps employed on laboratory exhaust stacks?	___	___	___	___

N/A - Not Applicable

WORKPLACE HEALTH PROTECTION PROGRAM ELEMENT CHECKLIST ON LABORATORY OPERATIONS

		Yes	No	Don't Know	N/A
60.	Is exhaust fan rotational direction proper for each system?	___	___	___	___
61.	Are fan discharge points sufficiently far from air intakes to prevent reentry of exhaust?	___	___	___	___
62.	Are heat exchangers cleaned after a shutdown period to prevent the growth and dispersal of biological organisms (e.g., Legionella) into the laboratory air distribution system?	___	___	___	___
63.	Are compressed gas and utility lines (air, nitrogen, etc.) identified?	___	___	___	___
64.	Are compressed gas cylinders:				
	. Properly identified?	___	___	___	___
	. Supported properly?	___	___	___	___
	. Stored in an adequately ventilated area?	___	___	___	___
	. Separated to keep oxidizing gases away from flammable gases?	___	___	___	___
	. Provided with a proper regulator?	___	___	___	___
	. Provided with a proper transfer line?	___	___	___	___
	. The proper size for the application?	___	___	___	___
	. Kept out of the laboratory?	___	___	___	___
	. Stored at sufficient distance (e.g., 75 feet or more) from air supply intakes?	___	___	___	___
65.	Are gas storage rooms provided with a separate ventilation system?	___	___	___	___

N/A - Not Applicable

WORKPLACE HEALTH PROTECTION PROGRAM ELEMENT CHECKLIST ON LABORATORY OPERATIONS

	Yes	No	Don't Know	N/A
66. Is laboratory illumination adequate:				
. Offices [50 footcandles (fc) - 500 lux]?	___	___	___	___
. Laboratory benches (70 fc - 700 lux)?	___	___	___	___
. Hoods (50 fc - 500 lux)?	___	___	___	___
. Aisles (30 fc - 300 lux)?	___	___	___	___
67. Is emergency lighting provided in laboratory areas?	___	___	___	___
68. Is emergency lighting tested periodically to ensure operability and adequacy?	___	___	___	___
69. Are vacuum pumps properly exhausted?	___	___	___	___
70. Are ionizing radiation sources properly identified?	___	___	___	___
71. Are sealed sources of ionizing radiation periodically leak tested at appropriate intervals (e.g., every 6 months)?	___	___	___	___
72. Is laboratory employee exposure to ionizing radiation monitored (when appropriate)?	___	___	___	___
73. Are laboratory personnel periodically monitored to determine their exposure to chemical substances?	___	___	___	___
74. Are employees who meet appropriate considerations included in applicable medical surveillance programs (e.g., respirator, radiation, lead, etc.)?	___	___	___	___
75. Are monitoring results properly maintained?	___	___	___	___

N/A - Not Applicable

WORKPLACE HEALTH PROTECTION PROGRAM ELEMENT CHECKLIST ON LABORATORY OPERATIONS

		Yes	No	Don't Know	N/A
76.	Is eating prohibited in the laboratory?	___	___	___	___
77.	Is there a separate lunch room at the laboratory?	___	___	___	___
78.	Are chemical storage shelves provided with a spill lip?	___	___	___	___
79.	Is the chemical storage area adequately ventilated?	___	___	___	___
80.	Are chemicals stored properly:				
	. Samples?	___	___	___	___
	. Retain samples?	___	___	___	___
	. Reagents?	___	___	___	___
	. Compressed gases?	___	___	___	___
81.	Is floor storage of chemicals prohibited?	___	___	___	___
82.	Is a bottle carrier used to transport hazardous chemicals?	___	___	___	___
83.	Are stocks of chemicals checked periodically to determine:				
	. Leakage?	___	___	___	___
	. Corrosion of containers?	___	___	___	___
	. Evaporation?	___	___	___	___
	. Expiration date?	___	___	___	___
	. Incompatibility potential?	___	___	___	___
84.	Are samples effectively disposed (e.g., in a hood sink) in order to reduce exposure to the material?	___	___	___	___
85.	Is there an on-going effort to reduce waste discharge from laboratory operations?	___	___	___	___

N/A - Not Applicable

WORKPLACE HEALTH PROTECTION PROGRAM ELEMENT
ON
MECHANICAL SHOPS

WORKPLACE HEALTH PROTECTION PROGRAM ELEMENT ON MECHANICAL SHOPS

OBJECTIVE

To identify potential health hazards associated with maintenance work in the mechanical shops and control methods for reducing exposure to these.

BACKGROUND

The workforce in petroleum and petrochemical operations is generally divided equally between operations and maintenance employees. Maintenance personnel are assigned work at outside locations, as well as in a building referred to as the shops. Mechanical craft personnel perform similar tasks in both locations. In most facilities, for example, welders, pipefitters, machinists, electricians, carpenters, and other crafts are assigned work in the field or the shops. The likelihood for these maintenance personnel to be exposed to hazardous chemical, physical or biological agents is similar for both shop and field work, but, the characteristics of the exposures can be different. For example, the level of exposure to hydrocarbon vapors, acutely toxic gases (e.g., H_2S, SO_2, Cl_2, etc.), noise, heat, and other health hazards that maintenance personnel experience in field work is typically higher than occurs in shop operations. However, exposure periods are often longer for shop personnel than for their field counterpart. Thus, the exposure dose of both groups is similar during routine operations, particularly as relates to chronic acting agents. The exposure of field personnel can be considerably greater for repair/maintenance activities during turnaround periods.

CONSIDERATIONS

To determine the health concerns associated with work in the mechanical shops, it is necessary to identify the tasks that are performed; substances being used; how much of a substance of concern is used; frequency that a task is carried out; where exposure to the contaminant of concern may occur; and control measures in place, and their effectiveness for reducing exposure to airborne contaminants, physical agents, and other health hazards of concern. Information obtained in carrying out this type qualitative assessment can provide a basis for conducting quantitative exposure assessments of personnel working at various operations that are performed. Examples of some of these operations are shown in the accompanying table, along with a listing of contaminants of concern and methods typically employed to reduce exposure to acceptable levels. Other workplace health protection issues associated with shop work include:

- Use of protective equipment may be necessary to reduce exposure of personnel to some shop hazards. This may include eye protection, face shield, gloves, apron, hearing protection, respirator, welding shield, abrasive blasting helmet, and others.

- Contaminants generated during welding, brazing and lead burning should be removed from the work area employing well designed mechanical exhaust ventilation systems. The effectiveness of these exhaust systems should be evaluated periodically and measures taken to improve performance, when warranted.

- Some repair/maintenance operations conducted in the shop can generate significant noise energy. For example, grinding, submerged arc welding, plasma arc cutting, safety valve testing, some machining operations, sheet metal fabrication and other work can result in employees being excessively exposed to noise. Personnel involved in such work (and others nearby) should be provided, and required to wear, suitable hearing protection if their exposure to noise is excessive. Engineering controls should be installed, where feasible.

- Operations carried out in the shops require good illumination. Supplementary (direct) lighting may be necessary at some tasks. Lighting should be evaluated periodically and improvements made, when warranted. A lighting maintenance program should be implemented to ensure that recommended illumination levels are established and maintained.

- Equipment to be taken to the shop for repair/maintenance should be purged with a low hazard material, flushed with water, then drained before transfer from the field. This reduces the potential for exposure of personnel (riggers, shop personnel) to the potentially hazardous material handled by the equipment when in service. When appropriate, equipment should be tagged to indicate it may be contaminated with a hazardous material.

- Equipment brought to the shop for repair/maintenance may have asbestos insulation on it. As much of the asbestos-containing material (ACM) as possible should be removed before release from the unit/area for transfer to the shop. If effective ACM removal cannot be achieved in the field, the equipment should be bagged and tagged to identify the presence of the contamination. If this occurs, special procedures/work practices appropriate for work with ACM, (establish a control area, limit access, require use of personal protective equipment by personnel doing repair/maintenance, etc.) will have to be followed by personnel performing the work in the shop.

- During repair procedures, exhaust gases from internal combustion engines should be directed outside the shop through flexible duct connected to the exhaust pipe. Crushed or leaking ductwork should be replaced to facilitate effective exhaust discharge.

- Cutting oils and cutting fluids may produce dermatitis among personnel who contact these materials. Appropriate protective equipment and personal hygiene facilities must be available and used by personnel to minimize skin contact and remove contamination should it occur. Manufacturer's instructions should be followed regarding the addition of bactericidal agents and changing the fluids.

- Mercury is used in some gauges. Mercury spills may occur when pressure testing them. If mercury is spilled it should be cleaned up promptly to prevent exposure to its vapor.

- Abrasive cleaning operations should be carried out in a manner which reduces exposure to abrasive and pigment dust, as well as noise. Effective mechanical exhaust ventilation is necessary for abrasive cleaning cabinets and cleaning rooms, and appropriate protective equipment (respirator, hearing protection, gloves, etc.) worn, depending on the associated exposure hazards.

- Degreasing/parts-cleaning containers located in the shop area should be provided with a low hazard solvent and kept covered when not in use. Contact with the liquid should be minimized and it should be changed when it becomes contaminated.

- Brazing rods/wire, typically referred to as silver solder, may contain cadmium. Work with the cadmium containing materials should be performed at an exhaust ventilated hood. If practical, use of cadmium containing brazing rods/wire should be prohibited.

- Fluorescent tubes removed from light fixtures are often brought to the shop for crushing in a tube buster. Broken glass, phosphor and mercury droplets are collected in a drum in this procedure. The waste drum should have a tight fitting lid and be stored outside the shop building until ready for disposal in an approved disposal facility. Use of respiratory protection is warranted when a large number of tubes are to be broken.

- Ballasts which contain PCBs should be properly disposed if removed from light fixtures or when light fixtures are to be discarded.

- Radiography is periodically carried out in the shop facility. If possible, this should be done during periods of non-occupancy (i.e., lunch time, evening). The Radiation Safety Officer should be contacted to review the job before it is performed to assure the work will be carried out in a satisfactory manner.

- Equipment contaminated with, or suspected to be contaminated with naturally occurring radioactive material (NORM), should be appropriately tagged to identify this potential before taken to the shop for repair. The equipment should be drained, flushed, and allowed to set idle for several hours (more than four) before transfer to the shop. When opened, residual scale should be retained for disposal as required by regulation (e.g., as low level radioactive waste). If personnel are to perform repair/maintenance work which will generate airborne particulates, then respiratory protection should be worn and exhaust ventilation used. Surface contamination should be cleaned up and appropriately disposed.

- All insulated equipment should be thoroughly decontaminated before transfer to the shop. As a precaution, it should be tagged to identify potential fiber contamination. When warranted, a bulk sample should be obtained and analyzed to determine if asbestos is present. Shop personnel must handle the equipment in a manner which prevents area contamination and exposure to fibers that may be released during repair/maintenance procedures. Appropriate respiratory protective equipment should be used, when warranted.

MECHANICAL SHOP EXPOSURE CONCERNS AND CONTROL

OPERATION	EXPOSURE CONCERNS	EXPOSURE CONTROLS
Abrasive Cleaning	Dusts (metal, pigment, silica, abrasive, etc.), fibers, noise, heat stress, vibration, cumulative trauma	Mechanical exhaust ventilation, respiratory protection, hearing protection, general ventilation, vibration dampening, personal hygiene
Degreasing	Solvent vapors, dermatitis	Water cooling jacket, exhaust ventilation, tight cover, limitation on rate of introduction/removal of items into/from cleaner, personal protective equipment
Electric Motor Repair	Solvent vapors, metal dust	General ventilation, personal hygiene, personal protective equipment
Gasket Cutting	Asbestos fibers	General ventilation, respiratory protection, housekeeping, replacement of material
Grinding	Dusts, noise, vibration, cumulative trauma	General ventilation, properly designed equipment, hearing protection, gloves
Instrument Repair	Solvent vapors, mercury vapor, solder fumes (metal/flux)	Mechanical exhaust ventilation, housekeeping, prompt cleanup of spilled mercury

MECHANICAL SHOP EXPOSURE CONCERNS AND CONTROL

OPERATION	EXPOSURE CONCERNS	EXPOSURE CONTROLS
Lead Burning	Lead fume	General ventilation for infrequent short periods up to one hour and less than once per week. If done more frequently, or for longer periods, perform work at a mechanical exhaust ventilated hood
Machining:		
Cutting Fluids	Mist, bacterial contamination, metal dust, dermatitis	Mechanical exhaust ventilation, bactericidal agent, personal hygiene, periodic change of oil, personal protective equipment
Cutting Oils	Mist, bacterial contamination, metal dust, dermatitis	Mechanical exhaust ventilation, bactericidal agent, personal hygiene, periodic change of oil, personal protective equipment
Metallizing	Metal fumes, dust, noise	Mechanical exhaust ventilated hood, hearing protection
Painting	Pigment mist, solvent vapor, dermatitis	Mechanical exhaust ventilated hood for spray painting, good ventilation for brush painting, personal protective equipment
Pipefitting	Metal fumes, dust, fibers, noise	General ventilation, hearing protection, other personal protective equipment, specific asbestos control procedures

MECHANICAL SHOP EXPOSURE CONCERNS AND CONTROL

OPERATION	EXPOSURE CONCERNS	EXPOSURE CONTROLS
Pump Maintenance/Repair	Liquids, vapors, gases, noise, NORM, fibers	General ventilation, personal hygiene, hearing protection, housekeeping, respiratory protection, other personal protective equipment, restrict access
Safety Valve Testing	Noise, gases, vapors	General ventilation, hearing protection, system to alert others of impending noise generation
Sheet Metal Work	Fumes (metal, flux and rosin), noise	General ventilation (if brazing is done frequently use mechanical exhaust ventilated hood), hearing protection
Vehicle Repair	Carbon monoxide, asbestos fibers, noise	General ventilation, use of flexible duct to discharge exhaust emissions outside, vacuum brake drums, wet wash to remove fibers after wheel is pulled or provide an exhaust ventilated enclosure, hearing protection, other personal protective equipment
Welding	Welding fumes (metal and flux), UV/IR radiation, ozone, carbon monoxide, oxides of nitrogen	Mechanical exhaust ventilated hood, general ventilation, eye/skin protection, welding screens, hearing protection

WORKPLACE HEALTH PROTECTION PROGRAM ELEMENT CHECKLIST ON MECHANICAL SHOPS

		Yes	No	Don't Know	N/A
1.	Is the shop clean and orderly?	___	___	___	___
2.	Are there adequate warning signs posted to identify potential hazard areas and the need for use of protective equipment (e.g., eye protection, hearing protection, etc.)?	___	___	___	___
3.	Has a complete listing been developed of materials (e.g., chemicals, paints, welding rods, etc.) in the shops?	___	___	___	___
4.	Have shop personnel been provided hazard awareness training?	___	___	___	___
5.	Are Material Safety Data Sheets available for all potentially hazardous materials to which shop personnel may be exposed?	___	___	___	___
6.	Have shop personnel been provided training on the proper procedures/ work practices appropriate for materials in use, or to which exposure can occur?	___	___	___	___
7.	Is appropriate personal protective equipment readily available to personnel?	___	___	___	___
8.	Have shop employees been trained in the proper use of personal protective equipment?	___	___	___	___
9.	Is personal protective equipment properly used?	___	___	___	___
10.	Have personnel who may be required to use respiratory protection been medically approved?	___	___	___	___
11.	Are there suitable washing facilities in the shop for personnel to practice good personal hygiene?	___	___	___	___

N/A - Not Applicable

WORKPLACE HEALTH PROTECTION PROGRAM ELEMENT CHECKLIST ON MECHANICAL SHOPS

		Yes	No	Don't Know	N/A
12.	Is drinking water available within a reasonable distance for shop personnel?	___	___	___	___
13.	Are adequate eating facilities provided for shop personnel?	___	___	___	___
14.	Are area trash receptacles provided with covers?	___	___	___	___
15.	Are vending machines clean and adequately serviced to ensure fresh, wholesome products?	___	___	___	___
16.	Has dermatitis, due to work related cause, been prevented among shop personnel?	___	___	___	___
17.	Are bactericidal agents added to cutting oils/fluids?	___	___	___	___
18.	Are cutting oils/fluids changed periodically?	___	___	___	___
19.	Have exhaust ventilation systems in operation in the shop been evaluated with respect to effectiveness within the past year?	___	___	___	___
20.	Are exhaust ventilation systems in operation in the shop effective for contamination control:				
	. Abrasive cleaning?	___	___	___	___
	. Brazing operations?	___	___	___	___
	. Degreasing?	___	___	___	___
	. Lead burning?	___	___	___	___
	. Metallizing?	___	___	___	___
	. Soldering?	___	___	___	___
	. Spray painting?	___	___	___	___
	. Welding operations?	___	___	___	___

N/A - Not Applicable

WORKPLACE HEALTH PROTECTION PROGRAM ELEMENT CHECKLIST ON MECHANICAL SHOPS

	Yes	No	Don't Know	N/A
21. Is general ventilation adequate in the shop to prevent build-up of contamination (CO, welding fumes, etc.)?	___	___	___	___
22. Are paints which contain lead or chrome excluded from shop use?	___	___	___	___
23. Are silica type abrasives excluded from use in abrasive cleaning operations?	___	___	___	___
24. Has cadmium containing silver solder been eliminated from use in brazing operations?	___	___	___	___
25. Is screening around welding areas adequate to prevent exposure to associated bright light in surrounding areas?	___	___	___	___
26. If shop personnel are required to enter confined spaces, have appropriate procedures been instituted and communicated (e.g., via training) to those involved?	___	___	___	___
27. Are solvents effectively used in electric motor cleaning operations?	___	___	___	___
28. Are compressed gas cylinders properly identified, stored and used?	___	___	___	___
29. If a fluorescent light tube buster is used, have the potential hazards associated with its use been communicated to those who operate it?	___	___	___	___

N/A - Not Applicable

WORKPLACE HEALTH PROTECTION PROGRAM ELEMENT CHECKLIST ON MECHANICAL SHOPS

		Yes	No	Don't Know	N/A
30.	Is the fluorescent light tube buster scrap drum provided with a tight fitting lid and stored in an unoccupied area?	___	___	___	___
31.	Is the noise exposure to shop personnel periodically evaluated?	___	___	___	___
32.	Is personal hearing protection readily available to shop personnel?	___	___	___	___
33.	Is illumination periodically evaluated in the mechanical shops?	___	___	___	___
34.	Is there an effective lighting maintenance program in place for the mechanical shops?	___	___	___	___
35.	If asbestos containing gasket material is cut in the shop:				
	. Has the potential hazard been communicated to employees?	___	___	___	___
	. Is only manual cutting performed?	___	___	___	___
	. Is the cutting table wet wiped after use to remove contamination?	___	___	___	___
	. Are respirators used during cutting?	___	___	___	___
36.	Is housekeeping satisfactory in the gasket cutting area?	___	___	___	___
37.	Is radiography that is done in the shop limited to off-hour (lunch time, evening, etc.) periods?	___	___	___	___
38.	Is the Radiation Safety Officer contacted to review radiographic work that is to be performed in the shop?	___	___	___	___

N/A - Not Applicable

WORKPLACE HEALTH PROTECTION PROGRAM ELEMENT CHECKLIST ON MECHANICAL SHOPS

	Yes	No	Don't Know	N/A
39. Are NORM contaminated pumps/equipment handled in a manner which:				
. Prevents exposure to the NORM?	___	___	___	___
. Prevents contamination of the area by NORM?	___	___	___	___
. Provides for collection and proper disposal of the NORM?	___	___	___	___
40. Are potentially contaminated NORM and asbestos sources identified when transferred to the shop?	___	___	___	___
41. Are solvent degreasing containers kept properly closed?	___	___	___	___
42. Are shop personnel informed of potential health hazards when temporarily assigned to field work during turnarounds?	___	___	___	___

N/A - Not Applicable

WORKPLACE HEALTH PROTECTION PROGRAM ELEMENT
ON
PACKAGING

WORKPLACE HEALTH PROTECTION PROGRAM ELEMENT ON PACKAGING

OBJECTIVE

To provide information on the workplace health protection issues associated with the operations carried out at petroleum/petrochemical packaging facilities and identify applicable exposure control considerations.

BACKGROUND:

Many petroleum and petrochemical products are packaged in various size containers for transfer to industrial and retail markets. Included are lubes, solvents, fuels, greases, oil dispersants, cutting oils/fluids, etc. Exposure to these materials can occur during packaging operations and appropriate controls are needed to minimize the potential for skin/eye contact or inhalation of the vapors or mists released during operations. Noise may be a concern, particularly when metal containers are used.

CONSIDERATIONS

Appropriate procedures/work practices and engineering controls need to be implemented to ensure that exposure to the components of the products being packaged is not excessive. Material safety data sheets (MSDSs) must be available to personnel as a source of information regarding the hazards associated with the products being handled and the control measures and protective equipment needs for preventing the occurrence of adverse health effects. Raw materials and products must be properly labeled. Recommended protective equipment must be available, personnel trained in its use, and operations identified where its use is indicated. Engineering controls should be provided, where necessary, to reduce the release or generation of hazardous agents and the subsequent exposure of personnel to them. This may involve providing noise controls on canning lines; eliminating noise emissions associated with compressed air releases; installing local exhaust ventilation at drum filling stations, lube blending tanks, grease kettles, and spray painting booths; as well as providing good general ventilation throughout the facility.

Internal combustion engine powered equipment should be adequately maintained to minimize emissions of hazardous combustion products (e.g., carbon monoxide, aldehydes, oxides of nitrogen, etc.,). Good general ventilation is necessary to prevent a build up of these contaminants in work areas. Appropriate personal protective equipment should be worn when maintaining/charging batteries in well ventilated areas.

Deluge showers/eyewash fountains should be provided, as warranted. These should be effectively identified, properly inspected and maintained, and personnel trained to use them in emergency situations.

Good housekeeping practices should be established and adhered to. Adequate personal hygiene facilities should be available to employees.

Ventilation systems should be evaluated periodically to ensure they continue to provide effective contamination control. Documentation of system performance should be maintained.

Illumination should be maintained at acceptable levels. To accomplish this, a lighting maintenance program is necessary due to the soiling and vibration effects of operations on lamps/fixtures.

REFERENCE

Chang, Shaw-Nong, "A Modified Ventilation Slot Hood for Chemical Drumming" American Industrial Hygiene Association Journal, July 1988, pp. 367-369.

**WORKPLACE HEALTH PROTECTION PROGRAM ELEMENT CHECKLIST
ON
PACKAGING**

		Yes	No	Don't Know	N/A
1.	Are packaging personnel aware of the health hazards associated with the materials being handled?	___	___	___	___
2.	Are material safety data sheets available to packaging personnel on the materials being handled?	___	___	___	___
3.	Are containers appropriately labeled to identify the health hazards associated with materials being handled?	___	___	___	___
4.	Is personal protective equipment (i.e. for the skin, hands and eyes) available and worn by personnel where contact with the materials being handled is possible?	___	___	___	___
5.	Have personnel been trained in the proper use of protective equipment?	___	___	___	___
6.	Have measures been taken to reduce noise generation/exposure associated with:				
	. Drum receipt?	___	___	___	___
	. Can filling?	___	___	___	___
	. Compressed air releases?	___	___	___	___
	. Steam releases?	___	___	___	___
7.	Are signs posted, where appropriate, indicating the need to wear hearing protection?	___	___	___	___
8.	Is hearing protection used, if warranted, at				
	. Filling lines?	___	___	___	___
	. Drum reclaiming?	___	___	___	___
	. Other operations? (__________)	___	___	___	___
9.	Is mechanical exhaust ventilation adequate at the:				
	. Drum painting spray booth?	___	___	___	___
	. Lube blending tanks?	___	___	___	___
	. Grease kettles?	___	___	___	___
	. Drum filling position?	___	___	___	___

N/A - Not Applicable

WORKPLACE HEALTH PROTECTION PROGRAM ELEMENT CHECKLIST ON PACKAGING

		Yes	No	Don't Know	N/A
10.	Is general ventilation adequate to prevent excessive exposure to solvent vapors during housekeeping work (cleaning equipment/floors with a solvent)?				
11.	Are mist releases minimized in association with the blow-down of product transfer lines?				
12.	Is personal protective equipment use adequate at hose manifold operations to prevent skin/foot/hand contamination?				
13.	Is appropriate protective equipment worn when charging/maintaining batteries on mobile equipment?				
14.	Is general ventilation adequate to minimize CO exposure associated with the operation of internal combustion engine powered equipment:				
	. Floor/sweeping?				
	. Transfer of packaged goods to storage?				
	. Loading trailers/railcars?				
15.	Are radioactive gauges (liquid level type) on canning lines:				
	. Registered, when appropriate?				
	. Appropriately identified?				
	. Periodically surveyed?				
	. Periodically wipe tested?				
16.	Are non-metal containers used in product packaging:				
	. < One liter?				
	. One liter?				
	. Five liter?				
	. Twenty liter?				
	. Drums?				

N/A - Not Applicable

WORKPLACE HEALTH PROTECTION PROGRAM ELEMENT CHECKLIST ON PACKAGING

	Yes	No	Don't Know	N/A
17. Is internal combustion engine powered equipment adequately maintained to minimize contaminant (CO, NOX, etc.) emissions?	___	___	___	___
18. Are deluge showers/eyewash fountains available to personnel, where appropriate?	___	___	___	___
19. Is illumination adequate throughout the facility?	___	___	___	___
20. Is a periodic lighting maintenance program in effect?	___	___	___	___
21. Is housekeeping adequate throughout the facility?	___	___	___	___
22. Are personal hygiene facilities adequate?	___	___	___	___
23. Do personnel practice good personal hygiene?	___	___	___	___

N/A - Not Applicable

WORKPLACE HEALTH PROTECTION PROGRAM ELEMENT
ON
PROCESS AND PRODUCT SAMPLING OPERATIONS

WORKPLACE HEALTH PROTECTION PROGRAM ELEMENT ON PROCESS AND PRODUCT SAMPLING OPERATIONS

OBJECTIVE

To provide information on process and product sampling in order to minimize the potential for employee exposure to the sampled material.

BACKGROUND

Process and product sampling procedures are carried out routinely in petroleum operations and petrochemical plants. The objective is to obtain a representative aliquot of the materials being used or produced. For example, raw feed, intermediates, catalyst, product, and other substances are removed from process, storage, or transport equipment for subsequent analysis (physical or chemical) to determine whether the material meets specifications. The information obtained in the analysis is used to make adjustments to the process to improve quality, increase yield, or for other reasons.

Samples may be obtained by manual or automatic methods. Both can result in a release of the material being sampled. The consequence of such a release may be the exposure of an individual to the substance being sampled. From a workplace health protection and environmental conservation standpoint the objective is to collect such samples with minimal employee contact and release to the air.

Other personnel may be exposed to, or come into contact with the sampled material. This may occur during analysis and disposal of the material, as well as during sample container cleaning and the maintenance of sample points (valves, connections, etc.).

CONSIDERATIONS

Materials that may be sampled include liquids, gases, and solids. Some may be highly flammable, combustible, and cold or hot. Others could be corrosive or irritating to the skin, eyes or respiratory tract if exposure or contact occurs. A few may have carcinogenic or reproductive health effects. Personnel need to be aware of the hazardous properties of the materials being sampled and the procedures to follow to obtain the sample without undue risk. It should be emphasized that there will always be some potential for exposure to a material while collecting a sample of it. Every effort should be made to minimize the likelihood for such an exposure to occur.

Effective procedures, proper containers, and appropriate personal protective equipment must be used to reduce exposure potential during sampling. In addition, sample points must be properly designed and maintained.

It is essential that sampling be done only by personnel who have been trained in the proper procedures, provided the proper protective equipment, and made aware of the hazards associated with the materials being sampled. They must be trained in the use of the protective equipment they use and the action to take if exposure or contact with the material occurs. Where warranted, a properly operating deluge shower/eyewash fountain should be located near (within 50 feet) the collection point. In addition, personal hygiene facilities (sink with soap and running water, towels, etc.) should be available to personnel, when appropriate.

Various types of containers are used in sampling. The type to use is dictated by the properties of the material, analytical needs, as well as temperature and pressure considerations. The proper type must be provided and used.

Most samples are taken to a laboratory for analysis. A sample storage area should be established there to receive and store the samples until analyzed. Accepted practices should be established and adhered to in the handling and disposal of samples. Some laboratories have a special sink that is piped to a sump for sample disposal. Disposal by dumping the material onto the ground or venting into the air can result in exposure to the material, as well as environmental contamination. Accepted methods should be established.

Following is a listing of some of the measures that should be implemented to minimize the potential for employee exposure to hazardous materials during sampling operations:

- Install accessible sample points that are properly designed and maintained for minimizing the release of material.
- Provide personnel training in proper sampling procedures.
- Provide the proper type sample container for the material being sampled.
- Provide training to personnel regarding the potential hazards associated with the materials being sampled.
- Ensure that personnel know what action to take if a release occurs, or if they are exposed to the material.
- Provide personnel the proper protective equipment for sample collection.
- Train personnel in the proper use of the protective equipment.
- Establish special procedures for sampling hydrogen sulfide-containing materials.

- Minimize the release of material during sample line flushing and purging.
- Prohibit the flushing of liquid samples from the sample point to the ground.
- Prohibit the disposal of sample container rinsings onto the ground.
- Provide closed-loop sample systems, where appropriate, to eliminate the release of sampled material.
- Keep sample volumes as small as practical.
- Close sample valves tightly after sample collection.
- Report leaking sample valves/connections promptly for maintenance.
- Provide appropriate outage for sample containers and close them tightly (after cooling) to prevent leakage.
- Label sample containers with appropriate identification of contents and the hazards of the material (e.g., toxic, corrosive, etc.).
- Provide for the safe transport and storage of samples.
- Develop and implement an acceptable method for discarding samples and retains.

WORKPLACE HEALTH PROTECTION PROGRAM ELEMENT CHECKLIST ON PROCESS AND PRODUCT SAMPLING OPERATIONS

	Yes	No	Don't Know	N/A
1. Have personnel who are involved in process/product sampling been trained regarding:				
. Proper sampling procedures?	___	___	___	___
. Hazards of the sampled materials?	___	___	___	___
. Use of protective equipment?	___	___	___	___
. Measures to take if contact/exposure occurs?	___	___	___	___
2. Are there written procedures for collecting samples from equipment?	___	___	___	___
3. Are the proper type containers used for sampling:				
. Solids?	___	___	___	___
. Liquids?	___	___	___	___
. Gases?	___	___	___	___
. Hot/cold materials?	___	___	___	___
4. Is appropriate protective equipment available and used by personnel when collecting samples?	___	___	___	___
5. Are personnel who are involved in sampling aware of the location of emergency equipment?	___	___	___	___
6. Is the proper protective equipment worn during sampling to protect:				
. Eyes?	___	___	___	___
. Face?	___	___	___	___
. Hands?	___	___	___	___
. Body?	___	___	___	___
. Feet?	___	___	___	___
. Respiratory system?	___	___	___	___
7. Is the deluge shower/eyewash fountain:				
. Properly designed?	___	___	___	___
. Effectively identified?	___	___	___	___
. Readily accessible?	___	___	___	___
. Provided with potable water?	___	___	___	___
. Provided with proper pressure and flow?	___	___	___	___
. Periodically tested and results recorded?	___	___	___	___

N/A - Not Applicable

WORKPLACE HEALTH PROTECTION PROGRAM ELEMENT CHECKLIST ON PROCESS AND PRODUCT SAMPLING OPERATIONS

		Yes	No	Don't Know	N/A
8.	During sample collection, do personnel avoid standing in a downwind position, as much as practical?	___	___	___	___
9.	Is sample line flushing/purging to grade prohibited during sample collection?	___	___	___	___
10.	Where appropriate, are closed loop sampling systems:				
	. In place?	___	___	___	___
	. Being considered?	___	___	___	___
11.	Is there an approved device available for unplugging sample lines?	___	___	___	___
12.	Is it possible to isolate sample points if a line breaks or a valve sticks in the open position?	___	___	___	___
13.	Are sampling system valves opened in the proper sequence to minimize auto-refrigeration of valve connections?	___	___	___	___
14.	Is appropriate protective equipment used during sampling of:				
	. Benzene containing streams?	___	___	___	___
	. Corrosive liquids?	___	___	___	___
	. Flue gas?	___	___	___	___
	. Hot catalyst?	___	___	___	___
	. Liquefied gas?	___	___	___	___
	. Streams containing H_2S?	___	___	___	___
	. High toxicity materials?	___	___	___	___
15.	Are sample valves effectively closed (to stop leakage) after sample collection?	___	___	___	___
16.	Are leaking sample points promptly repaired?	___	___	___	___

N/A - Not Applicable

WORKPLACE HEALTH PROTECTION PROGRAM ELEMENT CHECKLIST ON PROCESS AND PRODUCT SAMPLING OPERATIONS

		Yes	No	Don't Know	N/A
17.	Are hot sample containers adequately cooled before closing?	___	___	___	___
18.	When appropriate, is an adequate outage volume retained in sample containers (e.g., about 15 percent)?	___	___	___	___
19.	Are sample containers washed (when appropriate) to remove contamination before transfer to the laboratory?	___	___	___	___
20.	Is sample placed in a proper carrier (to prevent breakage, spillage, contact) for transport to the laboratory?	___	___	___	___
21.	Are liquid samples carried in an upright position?	___	___	___	___
22.	Is the sample transported outside the passenger section of a vehicle?	___	___	___	___
23.	Are samples prohibited from being takn into a control room, office, or other inside (non-laboratory) location?	___	___	___	___
24.	Is there an acceptable, designated sample placement area in the laboratory?	___	___	___	___
25.	Are laboratory samples disposed in an acceptable manner?	___	___	___	___
26.	Is the bottle-wash room housekeeping acceptable?	___	___	___	___
27.	Is the bottle-wash room sink acceptable for disposal of sampled materials?	___	___	___	___
28.	Is the bottle-wash room ventilation effective for removing vapors?	___	___	___	___

N/A - Not Applicable

WORKPLACE HEALTH PROTECTION PROGRAM ELEMENT CHECKLIST ON PROCESS AND PRODUCT SAMPLING OPERATIONS

	Yes	No	Don't Know	N/A
29. Is there a deluge shower/eyewash fountain in the:				
. Laboratory?	___	___	___	___
. Bottle-wash room?	___	___	___	___
. Sample retain facility?	___	___	___	___
30. Are potential hazards identified to maintenance personnel prior to their working on sampling systems?	___	___	___	___
31. Is required protective equipment provided to maintenance personnel prior to their working on sampling systems?	___	___	___	___
32. Do maintenance personnel wear appropriate protective equipment when working on sampling systems?	___	___	___	___
33. Is there an effective method and procedure for purging and flushing sample lines/valves/etc. before maintenance work is done?	___	___	___	___
34. Do maintenance personnel take appropriate measures to minimize release of materials during work at sample points?	___	___	___	___
35. Is sample point accessibility a consideration in process equipment design and installation?	___	___	___	___

N/A - Not Applicable

WORKPLACE HEALTH PROTECTION PROGRAM ELEMENT
ON
RECEIPT AND STORAGE OF MATERIALS

WORKPLACE HEALTH PROTECTION PROGRAM ELEMENT ON RECEIPT AND STORAGE OF MATERIALS

OBJECTIVE

To provide information useful for the development and implementation of an effective program which addresses the potential workplace health concerns associated with material receipt and storage, and identifies control procedures to effectively reduce exposure potential.

BACKGROUND

Awareness of the potential hazards associated with materials received, stored and used in a facility is necessary to reduce the risk to personnel who may be exposed to these substances. This can be achieved by obtaining and making available to personnel relevant information, such as material safety data sheets. This type information should be obtained from the suppliers of all products used. The sheets should be maintained in an orderly manner for reference or copying and use by personnel in the facility. It is good practice if new material requisitions are reviewed and approved (by industrial hygiene, safety and environmental control personnel) before the purchase order is issued. This can eliminate the introduction of high hazard materials into the facility with the potential for excessive exposure and the need for special handling, extra exposure controls, and costly disposal if not used. In addition, proper labeling by the supplier is essential for alerting personnel of the exposure concerns at the time of use of the substance.

Potentially exposed personnel should be included in a hazard awareness training program designed to convey information on the hazards associated with the materials received, stored and used. This will enable users to be aware of the measures they should take in the event of breakage, spillage or other type release that could result in their being exposed to, or coming into contact with the material.

Improper storage of materials can give rise to chemical incompatibilities, or the decomposition of some products with possible off-gassing. Storing items in alphabetical order is not good practice. J. T. Baker (Phillipsburg, NJ) provides information on chemical storage to prevent incompatibility problems. In addition, the U.S. Coast Guard has developed chemical incompatibility information. Adequate ventilation is essential to remove vapors and gases emitted by materials in storage. Temperature control is often needed to maintain the stability of some stored materials. It is advisable to date materials when received and require stock rotation to ensure the oldest material is used first.

CONSIDERATIONS

The employer is responsible for identifying the potential health hazards associated with the materials received, stored, used, and produced in operations. Relevant information must be obtained and conveyed to employees, along with information on procedures/work practices and control methods for reducing exposure risk. These include:

- Developing and maintaining a complete inventory of the materials received, stored, and used on site.
- Obtaining information (e.g., MSDSs) on the potential hazards associated with all materials received, stored, used, and produced on site.
- Developing and presenting a hazard communication program to all personnel who are potentially exposed to the hazardous materials. New hires should receive training before being exposed to the hazards in the workplace.
- Assuring that all hazardous material containers are properly labeled and stored, and that potentially exposed personnel are aware of the associated hazards.
- Providing personal hygiene facilities and appropriate personal protective equipment to effectively reduce exposure to the hazardous substances received and stored in order to enable personnel to take appropriate measures to reduce adverse health effects in the event exposure/contact occurs.
- Training personnel in the proper use of protective equipment that is appropriate for the health hazards they may encounter.
- Assuring that storage areas are properly designed for the materials received and stored, and that personnel adhere to proper procedures during handling.
- Establishing a procedure for the review of materials proposed for use on site so that potential health hazards are identified beforehand and that adequate precautions are in place for the safe receipt, storage, use, and disposal of the substances.
- Developing written emergency procedures to be followed in the event of a spill of a hazardous material.
- Assuring that emergency response personnel are appropriately trained for the situations they may

encounter in association with the receipt and storage of hazardous materials.

- Reviewing storage practices periodically to ensure that incompatibilities do not exist.
- Storing flammables, acids, and highly toxic substances in separate dedicated storage cabinets.
- Dating materials when received.
- Avoiding storage of chemicals on the floor and above eye level.

WORKPLACE HEALTH PROTECTION PROGRAM ELEMENT CHECKLIST ON RECEIPT AND STORAGE OF MATERIALS

		Yes	No	Don't Know	N/A
1.	Has a complete inventory been developed of the materials received, stored, used, and produced on site?	___	___	___	___
2.	Have material safety data sheets, or similar information, been obtained on all materials received, stored, used, or produced on site?	___	___	___	___
3.	Are material safety data sheets or other similar information complete, up-to-date and maintained in an organized manner and readily available?	___	___	___	___
4.	Are materials that are received properly labeled?	___	___	___	___
5.	Is there a procedure to date chemicals before placing them in storage?	___	___	___	___
6.	Is there an established practice of rotating stock so that older items are used first?	___	___	___	___
7.	Are materials in storage properly labeled?	___	___	___	___
8.	Are labels adequate for conveying hazard information to personnel?	___	___	___	___
9.	Are requisitioned materials reviewed with regard to workplace health hazard potential before purchase?	___	___	___	___
10.	Are material safety data sheets or similar information made available in work areas for users of materials?	___	___	___	___

N/A - Not Applicable

WORKPLACE HEALTH PROTECTION PROGRAM ELEMENT CHECKLIST ON RECEIPT AND STORAGE OF MATERIALS

		Yes	No	Don't Know	N/A
11.	Is there a formal, documented training program in place to inform potentially exposed personnel of the hazards associated with the receipt, storage, and use of hazardous materials to which they may be exposed?	___	___	___	___
12.	Are potentially exposed employees aware of the health hazards associated with the receipt and storage of all materials they may contact or be exposed to?	___	___	___	___
13.	Are samples of materials (e.g., chemicals, products, process streams, etc.) stored outside until transfer to the laboratory, retain facility, etc.	___	___	___	___
14.	Have personnel been trained in the procedures/work practices/ protective equipment to use for minimizing exposure to or contact with hazardous materials?	___	___	___	___
15.	Is appropriate personal protective equipment readily available to receiving personnel?	___	___	___	___
16.	Are emergency deluge showers/eyewash fountains available in the immediate vicinity (within 15 meters) of locations where contact with corrosives, irritants, or other hazardous substances may occur?	___	___	___	___
17.	Are deluge shower/eyewash fountain facilities in receiving/storage areas:				
	. Properly designed?	___	___	___	___
	. Readily accessible?	___	___	___	___
	. Adequately identified?	___	___	___	___
	. Operating properly?	___	___	___	___

N/A - Not Applicable

WORKPLACE HEALTH PROTECTION PROGRAM ELEMENT CHECKLIST ON RECEIPT AND STORAGE OF MATERIALS

	Yes	No	Don't Know	N/A
. Periodically checked to assure proper operation for emergency use?				
. Provided with potable water?				
. Provided with water at acceptable temperature and pressure?				
(Frequency of Testing?________________)				
18. Are adequate facilities available in material receiving and storage areas to enable personnel to practice good personal hygiene?				
19. Are materials properly stored to prevent the occurrence of incompatibilities?				
20. Are toxic materials properly received and stored in the warehouse or other location until delivery to, or pick up by the user/requisitioner?				
21. Do users of chemicals properly store and handle them?				
22. Are storage areas adequately ventilated?				
23. Are bottle carriers used for transporting hazardous materials?				
24. Is lighting adequate in storage areas?				
25. Are powered industrial trucks that are used in receipt/storage operations:				
. Battery powered?				
. Internal combustion type:				
- Diesel?				
- Propane or LPG?				
- Gasoline?				

N/A - Not Applicable

WORKPLACE HEALTH PROTECTION PROGRAM ELEMENT CHECKLIST ON RECEIPT AND STORAGE OF MATERIALS

		Yes	No	Don't Know	N/A
26.	Is the mobile equipment battery charging area provided with:				
	. Adequate ventilation?	___	___	___	___
	. Deluge shower/eyewash fountain nearby?	___	___	___	___
27.	Is fueling of internal combustion type industrial trucks carried out in an acceptable location relative to the potential for exposure to fuel vapors?	___	___	___	___
28.	Are powered industrial trucks and sweepers adequately maintained to minimize CO emissions?	___	___	___	___
29.	Is ventilation adequate to minimize the build-up of contaminants in areas in which internal combustion powered equipment is operated?	___	___	___	___
30.	Are compressed gases properly stored?	___	___	___	___
31.	Is video display equipment used by receiving personnel located to prevent glare?	___	___	___	___
32.	Is there a pan beneath receiving manifolds to contain spills, should they occur?	___	___	___	___
33.	If there is a spill of hazardous material in a receipt or storage area, is there an established procedure for containment, clean-up, and assessing the environment for reentry?	___	___	___	___

N/A - Not Applicable

WORKPLACE HEALTH PROTECTION PROGRAM ELEMENT CHECKLIST ON RECEIPT AND STORAGE OF MATERIALS

	Yes	No	Don't Know	N/A
34. Are spill response personnel provided:				
. Hazard communication training?	___	___	___	___
. Training in the use of personal protective equipment?	___	___	___	___
. Monitoring equipment to assess exposure risk?	___	___	___	___
. Medical examinations?	___	___	___	___
35. Is there a written procedure for the receipt and storage of radioactive materials?	___	___	___	___
36. Does receiving adhere to acceptable procedures for receipt of radioactive materials?	___	___	___	___
37. Has a review of the facility's receipt and storage practices (i.e., relative to adverse health effect potential) been carried out within the past 2 years?	___	___	___	___
38. Are vehicle operators aware of the hazards associated with materials being transported?	___	___	___	___
39. Is appropriate personal protective equipment available and used by personnel involved in the bulk receipt of hazardous materials?	___	___	___	___
40. Is there a written procedure on the action to take if an emergency arises during the receipt of a bulk material?	___	___	___	___
41. Is lighting adequate in areas where bulk products are received?	___	___	___	___

N/A - Not Applicable

WORKPLACE HEALTH PROTECTION PROGRAM ELEMENT
ON
TANK CLEANING

WORKPLACE HEALTH PROTECTION PROGRAM ELEMENT ON TANK CLEANING

OBJECTIVE

To provide information on workplace health protection considerations for tank cleaning operations.

BACKGROUND

In the petroleum and petrochemical industry, tanks of various sizes and shapes are used to store water, chemicals, crude oil, petroleum products, liquified gas, and other materials. Some have fixed roofs while others have floating (internal or external) roofs. Some may have no roof. Others may be cylindrical (in the vertical or horizontal mode) or spheroidal. Many may be lined on the inside, whereas the inside of others is bare metal. Some are used to store materials which are at significant pressure.

Materials of tank construction include aluminum, carbon or alloy steel, concrete, or fiberglass-reinforced plastic. Underground gasoline storage tanks presently being installed are generally double tanks to reduce the potential for leakage and the release of product to the surrounding soil.

There are a multitude of reasons for cleaning tanks that are in use. These include: for periodic inspection, maintenance, changing service, repairing a leak, to remove sludge, and others. The preparations to enter tanks and carry out an inspection, cleaning, or other work are generally the same. They involve emptying, blanking, water washing, vacuuming out as much sludge as possible, and ventilating the tank until the contaminant concentration is acceptable or sufficiently low to enable entry with respiratory protection use. Some tanks may be inerted (i.e., with an inert gas) to reduce the fire/explosion potential.

CONSIDERATIONS

A written generic tank cleaning guide should be developed and provided to tank cleaning contractors, in-house supervision, and other personnel involved with tank cleaning work. It should incorporate information relative to entry permit requirements; tank cleaning procedures; atmosphere testing requirements; the need for a standby person; area access restrictions; recommended ventilation and monitoring practices; equipment needs; special health concerns associated with some materials that may be present (e.g., hydrogen sulfide, organic lead compounds, benzene, etc.), and other information that will help ensure tank cleaning work is completed without adverse health effects occurring among personnel.

Contractors who may bid on tank cleaning work should be informed of the special requirements and potential health hazards associated with the cleaning operation when bids for the work are requested. The contractors should be required to submit written copies of their workplace health protection and/or safety program with bid submissions, along with a listing of hazardous materials to be brought on site for the tank cleaning work. These should be reviewed to ensure the contractor has addressed associated health concerns. It is not the intent to ascertain their compliance with applicable regulations in detail, but rather to assure they recognize the potential health hazards and will have controls in place that are adequate to prevent adverse health effects to their personnel, as well as to facility employees. For example, does the contractor have an effective written respiratory protection program which meets the need when implemented is the question to be addressed, not if it is in compliance with regulatory requirements. General information provided by the contractor, along with details on each workplace health protection program element, can serve as one basis for contractor selection. Results of these reviews should be documented.

A pre-cleaning meeting should be held with the tank cleaning supervisors. Entry procedures, the permit system, blanking and ventilation requirements, lock out/tag out procedures, etc., should be discussed. The tank should be inspected internally before the work is begun and a report prepared of conditions.

REFERENCES

"Cleaning Petroleum Storage Tanks", API Publication 2015, 3rd Edition (Washington, D.C.: American Petroleum Institute, September 1985).

"Guide for Controlling the Lead Hazard Associated with Tank Entry and Cleaning", API Publication 2015A, 2nd Edition (Washington, D.C.: American Petroleum Institute, June 1982).

"Cleaning Open-Top and Covered Floating Roof Tanks", API Publication 2015B (Washington, D.C.: American Petroleum Institute, August 1981).

"Atmospheric and Low-Pressure Storage Tanks", Chapter XIII, Guide for Inspection of Refining Equipment, Fourth Edition (Washington, D.C.: American Petroleum Institute, April 1981).

WORKPLACE HEALTH PROTECTION PROGRAM ELEMENT CHECKLIST ON TANK CLEANING

		Yes	No	Don't Know	N/A
1.	Are hydrocarbon liquids in a tank pumped down to the lowest level possible before a tank is opened?	___	___	___	___
2.	Are tanks blinded and blinds inspected before the tank is opened?	___	___	___	___
3.	Are hydrocarbon vapors exhausted from tanks through top manways to minimize vapor contamination around the tank?	___	___	___	___
4.	Is tank cleaning considered confined space work (when entry is required)?	___	___	___	___
5.	Are there written procedures for tank cleaning work?	___	___	___	___
6.	Are proper permits prepared and issued before tank cleaning work is started?	___	___	___	___
7.	Are the specific health hazards associated with a tank cleaning operation identified?	___	___	___	___
8.	Are the health hazards evaluated before entry/cleaning is permitted?	___	___	___	___
9.	Are monitoring results documented?	___	___	___	___
10.	Do requests for bids by contractors provide information to the contractors on potential health risks associated with the tank cleaning work to be done?	___	___	___	___
11.	Are contractor employee health protection issues reviewed before a contract is awarded:				
	. Hazard communication?	___	___	___	___
	. Personnel approved/trained for respirator use?	___	___	___	___

N/A - Not Applicable

WORKPLACE HEALTH PROTECTION PROGRAM ELEMENT CHECKLIST ON TANK CLEANING

	Yes	No	Don't Know	N/A
. Monitoring program addresses relevant issues?				
12. Are contractor workplace health protection program review findings documented as part of the contractor selection process?				
13. Are leaded gasoline tanks appropriately identified?				
14. Are tanks that contain sour material appropriately identified?				
15. Is a supervisor identified as being in charge of a tank cleaning operation?				
16. Is a tank pre-cleaning meeting held with contractor supervision before work is initiated?				
17. Are there provisions for stopping work if in-tank conditions change?				
18. After voiding a permit, is the cause investigated before a new permit is issued and work restarted?				
19. Are tank cleaning personnel:				
. Made aware of potential health hazards associated with this work?				
. Trained in tank cleaning procedures and work practices?				
. Trained with regard to use of personal protective equipment?				
. Aware of need for adhering to good personal hygiene practices?				
20. Is access to the tank area restricted until the tank vapor/gas level has been reduced to an acceptable level?				

N/A - Not Applicable

WORKPLACE HEALTH PROTECTION PROGRAM ELEMENT CHECKLIST ON TANK CLEANING

	Yes	No	Don't Know	N/A
21. Are pontoons and double decks on floating roof tanks vapor freed before tank work is initiated?	___	___	___	___
22. Is the space between the floating roof and the cone roof of an internal floating roof tank vapor freed before tank work is initiated?	___	___	___	___
23. Are leaded fuel storage tanks certified to be lead free before entry is permitted without respiratory protective equipment use?	___	___	___	___
24. Do personnel who enter leaded fuel storage tanks wear appropriate protective clothing?	___	___	___	___
25. Is the respiratory protective equipment that is specified for use in tank cleaning work based on monitoring results?	___	___	___	___
26. Is tank ventilation continued until all sludge is removed?	___	___	___	___
27. Is ventilation continued (where provided) while tank cleaning is in progress?	___	___	___	___
28. Is monitoring carried out during tank cleaning?				
. Area	___	___	___	___
- Continuously?	___	___	___	___
- Periodically?	___	___	___	___
. Personal	___	___	___	___
- Short-term?	___	___	___	___
- Long-term?	___	___	___	___
29. Is a stand-by person provided while tank cleaning progresses?	___	___	___	___
30. Is emergency rescue equipment available on site?	___	___	___	___

N/A - Not Applicable

WORKPLACE HEALTH PROTECTION PROGRAM ELEMENT CHECKLIST ON TANK CLEANING

		Yes	No	Don't Know	N/A
31.	Are cleaning operations periodically inspected to ensure appropriate procedures and work practices are being adhered to?	___	___	___	___
32.	Are personnel doing tank cleaning properly dressed to prevent contact with material in the tank?	___	___	___	___
33.	Is equipment (tools, personal protective equipment, etc.) effectively cleaned following completion of the tank cleaning operation?	___	___	___	___
34.	Are adequate facilities available to tank cleaning personnel to enable them to practice good personal hygiene?	___	___	___	___
35.	Do tank cleaning personnel follow good personal hygiene practices?	___	___	___	___
36.	Does a program exist to address acceptable disposal options for tank sludge?	___	___	___	___

N/A - Not Applicable

WORKPLACE HEALTH PROTECTION PROGRAM ELEMENT
ON
TURNAROUNDS

WORKPLACE HEALTH PROTECTION PROGRAM ELEMENT ON TURNAROUNDS

OBJECTIVE

To provide general information on workplace health protection issues associated with turnaround work and to identify a possible role for the industrial hygiene function prior to and during the turnaround.

BACKGROUND

A petroleum refinery, gas plant or petrochemical process unit cannot remain in operation indefinitely, and maintain optimum process and mechanical efficiency. Causes for this reduced efficiency include mechanical equipment maintenance needs, equipment deterioration (erosion, corrosion), catalyst degredation, accumulations of sludge in equipment, etc.

Maintenance of equipment is essential for efficient operations. Many routine maintenance tasks can be carried out during normal operations. However, for inspection and maintenance of major equipment, it is necessary to shut the process down periodically. Such a shutdown is typically referred to as a turnaround (T/A). These can be planned or unplanned, and either major or minor. Whichever the case, the process is shut down, the inspection, repair, and/or maintenance work completed, and the unit brought back on stream. The time involved can vary from a few days to a month or more.

Employees may, in some situations, be exposed to health hazards during preparation for shutdown, during the shutdown procedure, while turnaround work is being carried out, and during start-up. The facility personnel or contractors who do the work should be aware of the hazardous materials to which they may be exposed and of the appropriate measures to reduce exposures.

Tasks that are carried out before shutdown may include checking for leaks; determining pump maintenance needs; measuring metal erosion/corrosion; identifying areas where painting is needed; determining the orientation of trays in distillation towers; checking the condition of furnace burners, tubes, firebrick, etc; scaffold erection; evaluating the condition of insulation; inspecting rotary equipment; etc. Those doing this work may be exposed to various contaminants during the pre-shutdown period. Personnel need to be aware of the potential for exposure to these hazards, the procedures to follow to perform the task and the measures to take to minimize their exposure.

During the shutdown period, pumps, towers, drums, and other equipment to be worked on are isolated, drained, cleaned and vented before issuing a permit for additional work to be done. The potential for exposure to health hazards during this phase of a turnaround can be minimized or controlled if established procedures/work practices/confined space entry requirements are followed and the prescribed protective equipment is used.

During process start-up, the turnaround crew is typically away from the unit and operations personnel follow detailed written procedures to bring the unit back on line. This may take several days. Personnel may be exposed to hazardous materials and physical agents during this period. If equipment has not been closed tightly, leaks of feed, intermediates or product can result, and exposure to hydrocarbon vapors/gases and other contaminants can occur. Streams may be directed to a flare with resultant potential for exposure to noise, heat, acid gases, hydrocarbon vapors/gases, etc. Steam is often vented during start-up with accompanying high intensity noise. Process stream recycling, to establish operating conditions, can also give rise to noise generation.

The industrial hygiene function should be involved in turnaround planning in order to effectively provide support during the pre-shutdown, shutdown, turnaround and start-up periods. When facility employees are to be involved in the T/A work, hazard awareness and respirator training should be provided. The facility should assess exposures to the health hazards to which personnel may be exposed. When contractor personnel are to perform the turnaround work, the support provided by industrial hygiene would be in an advisory and oversight role to tell the contractor about potential hazards to which their personnel may be exposed and to periodically check to determine whether the contractor is adhering to good workplace health protection practices and complying with regulatory requirements.

CONSIDERATIONS

In some cases, it may be advisable for the industrial hygiene function to participate in the discussions regarding upcoming turnaround work so that plans can be developed to provide support related to the associated workplace health protection issues. Support should address the workplace health hazards associated with the pre-shutdown, shutdown, turnaround, and start-up activities. The scope of this support effort should be agreed upon before work is begun by employees or the contractor. It will depend on the unit to be turned around, the type of work to be carried out, the exposure potential associated with the tasks to be performed, and the toxic properties of contaminants to which people may be exposed. Other support activities could include:

- Reviewing contract bid proposals (relative to workplace health protection considerations)
- Identifying toxic/corrosive substances
- Identifying workplace health concerns associated with work to be done (i.e., radiography, insulation removal, leaded sludge handling, etc.)
- Deciding on the monitoring program to be carried out
- Identifying monitoring equipment needs/calibration/maintenance/etc.
- Identifying personal protective equipment needs/availability/training/etc.
- Deciding on ventilation equipment requirements and effective methods of use
- Conducting hazard awareness training
- Ensuring radiation source removal/storage/re-installation/wipe testing
- Reviewing procedures/work practices to be followed
- Establishing procedures (with supervisory personnel) for special jobs
- Ensuring appropriate emergency equipment is available (deluge shower/eyewash fountains/respiratory protective equipment/etc.) and ensuring its adequacy
- Reviewing previous turnaround experience

When there is a potential for personnel to be exposed to a toxic or corrosive material, appropriate protective equipment should be provided. Water pressure should be maintained at deluge showers/eyewash fountains and, if necessary, additional portable units located in appropriate locations for emergency use. Potable drinking water should be provided near each work location. Personnel operating impact tools should wear hearing protection, when appropriate. Good housekeeping practices should be established and maintained in each work area.

If employees are to perform the turnaround work, hazard communication and specific contaminant/physical agent awareness training should be provided to those personnel who will be involved and who may be exposed. When appropriate, employees should be provided respiratory protective equipment after medical approval and training in its use. If confined space work is to be performed, personnel should be provided appropriate training for that activity.

When it is considered appropriate, a monitoring program should be developed based on the contaminants to which personnel may be exposed, the work to be performed, and the associated exposure potential. General procedures, work practices, personal protective equipment usage, and available controls should be reviewed to determine their adequacy for minimizing exposure to the hazardous materials and physical agents associated with the tasks to be performed. If radiosotopes are to be used, the radiation safety officer should ensure that the source(s) is effectively shielded and the procedures to be followed are adequate to prevent excessive exposure to ionizing radiation during the work.

The industrial hygiene function, or other responsible group, should ensure that an adequate number of personal and portable monitors are calibrated and made available for assessing exposure risk during the turnaround. Where appropriate, personnel, such as gas testers, industrial hygiene monitors, or other designated individuals, should be trained in the use of this equipment and in interpreting results. The industrial hygiene function should provide support to these individuals, as necessary.

Arrangements should be made to provide additional signs, barricades, and protective equipment, such as deluge showers/eyewash fountains, respiratory protective equipment, breathing air, hearing and eye protection, hand and skin protection, personal hygiene facilities (wash basins, showers), etc. in the work area, as warranted. When indicated, such as for work at operations at which there may be exposure to lead, benzene, furfural, toluene or other contaminant for which a biological exposure index has been developed, biomonitoring and/or special medical evaluations could be considered. This decision is to be made based on discussions between management, the medical department and industrial hygiene personnel, or on regulatory requirements. Previous experience may provide information on which to make this determination.

If the turnaround work is to be done by contractors, then the role of the facility industrial hygiene function is typically one of oversight. It should be ensured that the contractor has effectively addressed the workplace health protection issues associated with the work; provided appropriate training to their personnel; is adhering to established procedures/work practices; is using the proper protective equipment; is adhering to bid specifications; and is in compliance with applicable regulations. Deficiencies noted are to be communicated to the contractor through established channels.

Those with industrial hygiene responsibility could visit the work site at least once a day, and more frequently if possible. The purpose of these visits is to provide advice/support to supervisory personnel, address concerns of employees (through their supervisor) as they arise, and observe the work in progress and, when appropriate, assess exposure risks and/or recommend control measures to minimize exposure to airborne contaminants/physical agents/skin contact, etc.

During the course of turnaround work, frequent checks could be made of emergency equipment to ensure it is in place and in proper working order. It is not uncommon for personnel (facility or contractor) to wrongfully use such equipment for routine tasks during a turnaround.

Breathing air quality should be checked frequently (more than usual) during a turnaround. Constant compressor operation may result in overheating with consequent breakdown of lubricating oil and resultant air contamination. It should be ensured that personnel do not jury-rig breathing air hose connectors to make them compatible with instrument or process air connections.

Portable monitoring equipment should be calibrated before use and checked periodically during the turnaround period to ensure it is working properly. If possible, each device should be calibrated once or twice a week, as necessary. The industrial hygiene function should review the response check and calibration procedures to ensure their effectiveness.

REFERENCES

"Guide for Inspection of Refinery Equipment - General Preliminary and Preparatory Work", 2nd Edition, Chapter III (Washington, D. C.: American Petroleum Institute, 1976).

"Guide for Inspection of Refinery Equipment - Preparation of Equipment for Safe Entry and Work", 3rd Edition, Chapter V (Washington, D. C.: American Petroleum Institute, 1978).

"Guide for Inspection of Refinery Equipment - Inspection for Accident Prevention", 3rd Edition, Chapter XIX (Washington, D. C.: American Petroleum Institute, 1984).

"Safe Maintenance Practices in Refineries", API Publication 1007 (Washington, D. C.: American Petroleum Institute, 1983).

WORKPLACE HEALTH PROTECTION PROGRAM ELEMENT CHECKLIST ON TURNAROUNDS

		Yes	No	Don't Know	N/A
1.	Does the industrial hygiene function participate in turnaround planning meetings?	___	___	___	___
2.	Does the industrial hygiene function develop a plan for providing support during turnaround activities?	___	___	___	___
3.	Does the industrial hygiene function provide support during turnaround operations:				
	. Advisory?	___	___	___	___
	. On-site?	___	___	___	___
	. Day shift?	___	___	___	___
	. 24 hours per day?	___	___	___	___
4.	Are there adequate industrial hygiene resources to respond to the workplace health protection issues associated with all turnaround work?	___	___	___	___
5.	Are outside industrial hygiene resources provided to assist facility personnel in supporting the workplace health protection program during turnarounds?	___	___	___	___
6.	Are periodic visits made to the turnaround work site to determine compliance with applicable requirements?	___	___	___	___
7.	Is a monitoring program developed and implemented when facility personnel are involved in the turnaround work?	___	___	___	___
8.	Are exposures evaluated during turnaround periods:				
	. Asbestos?	___	___	___	___
	. Hydrocarbon vapors?	___	___	___	___
	. Benzene?	___	___	___	___
	. Noise?	___	___	___	___
	. Ionizing radiation?	___	___	___	___

N/A - Not Applicable

WORKPLACE HEALTH PROTECTION PROGRAM ELEMENT CHECKLIST
ON
TURNAROUNDS

	Yes	No	Don't Know	N/A
. Welding fumes?	___	___	___	___
. Oxides of nitrogen?	___	___	___	___
. Carbon monoxide?	___	___	___	___
. Hydrogen sulfide?	___	___	___	___
. Catalyst dust?	___	___	___	___
. Other? (________________)	___	___	___	___
9. Is adequate equipment available for:				
. Area/source monitoring?	___	___	___	___
. Personal monitoring?	___	___	___	___
10. Is monitoring equipment:				
. Calibrated prior to and periodically during the turnaround?	___	___	___	___
. Provided a response check to assure effective operation before use?	___	___	___	___
11. Are calibration and response check procedures reviewed by the industrial hygiene function to ensure acceptability?	___	___	___	___
12. Are biomonitoring programs considered for personnel involved in turnaround work? (Circle if applicable: Benzene, Lead, Toluene, Other? (_____________)	___	___	___	___
13. Is potable water available in each work area during the turnaround period?	___	___	___	___
14. Are potential exposures determined for specific tasks and appropriate protective equipment provided personnel based on the results of such assessments?	___	___	___	___

N/A - Not Applicable

WORKPLACE HEALTH PROTECTION PROGRAM ELEMENT CHECKLIST ON TURNAROUNDS

		Yes	No	Don't Know	N/A
15.	Are spills cleaned up promptly when they occur?	___	___	___	___
16.	Is hearing protection worn during the operation of impact tools, when warranted?	___	___	___	___
17.	Is water pressure maintained at deluge showers/eyewash fountains during turnarounds?	___	___	___	___
18.	Is housekeeping acceptable in work areas during the turnaround period?	___	___	___	___
19.	Are facility personnel who are engaged in turnaround work provided:				
	. Hazard communication training?	___	___	___	___
	. Hearing protection use training (if applicable)?	___	___	___	___
	. Respirator use training (if applicable)?	___	___	___	___
	. Confined space work training (if applicable)?	___	___	___	___
20.	Are all sources of breathing air checked periodically during a turnaround?	___	___	___	___
21.	Does the industrial hygiene function provide workplace health hazard information to contractors involved in turnaround work on site?	___	___	___	___
22.	Does the industrial hygiene function provide oversite of contractor workplace health protection procedures/work practices and, where appropriate, communicate observed deficiencies to contractor supervisory personnel via established channels?	___	___	___	___

N/A - Not Applicable

**WORKPLACE HEALTH PROTECTION PROGRAM ELEMENT CHECKLIST
ON
TURNAROUNDS**

	Yes	No	Don't Know	N/A
23. Are adequate personal hygiene facilities available:				
. Toilets?	___	___	___	___
. Sinks?	___	___	___	___
. Showers?	___	___	___	___
. Change rooms?	___	___	___	___
24. Is lighting adequate for performing turnaround work?	___	___	___	___
25. Are the procedures for working with asbestos-containing materials acceptable?	___	___	___	___
26. Are wastes generated in turnaround work properly disposed:				
. Leaded sludge?	___	___	___	___
. Leaded paint pigments?	___	___	___	___
. Asbestos-containing materials?	___	___	___	___
. PCBs?	___	___	___	___

N/A - Not Applicable

WORKPLACE HEALTH PROTECTION PROGRAM ELEMENT
ON
VISUAL DISPLAY TERMINALS

WORKPLACE HEALTH PROTECTION PROGRAM ELEMENT ON VISUAL DISPLAY TERMINALS

OBJECTIVE

To provide information for assessing the workplace health concerns associated with visual display terminal (VDT) work stations and the considerations for their arrangement/design to minimize the potential for employees to experience discomfort or adverse health effect.

BACKGROUND

Over the past several years there has been a rapid growth in the use of visual display units (VDUs), or as they are more commonly known, visual display terminals (VDTs). Along with their use has been an increase in user complaints alleging a variety of adverse health effects. These have included headache, eye strain, musculoskeletal aches/pain, dermatitis, repetitive trauma disorders, cataracts, and others.

A number of studies have been conducted to evaluate the potential for adverse health effects to arise among VDT users. It has been reported that exposure to the electromagnetic radiations emanating from VDTs do not exceed national or international limits for continuous occupational exposure, and the potential for adverse health effects to arise as a result of such exposure during VDT use is considered negligible.

It has been determined that work at VDT terminals does not result in damage or permanent impairment to the visual system. The main eye effects that have been reported include eye fatigue and discomfort. These are avoidable through proper lighting, elimination of glare, providing rest breaks, and use of appropriate glasses. The musculoskeletal problems that have been reported are also preventable with good ergonomic design of the work station. Reported dermal effects (itching, erythema, rash, acne-like appearance of the skin) have not been confirmed and are still under investigation. The stress that some VDT operators reportedly experience, is not believed to be related to the operation of the VDT equipment.

Concern has been expressed regarding the potential for low level ionizing and non-ionizing radiations emitted by VDT equipment to adversely affect pregnancy outcomes. Investigations are on-going to determine such risk.

CONSIDERATIONS

The area in which a VDT is to be located should be painted with a non-reflective paint, and provided with overhead lighting that does not produce reflections on the VDT screen surface. Windows should be provided with adjustable coverings (curtains/blinds) to reduce bright light or glare on the screen. If screen reflections cannot be controlled through equipment positioning and adjustment, a polarized light filter can be mounted on the screen. Screen images should be free from flicker.

The workstation surface should be non-reflective and have adequate space to facilitate good housekeeping. The user's chair should be provided with a lumbar support and be easily adjustable from the seated position so that a pan height can be selected that is comfortable to the VDT operator while enabling proper knee angle (90-110°). There should be adequate room under the work surface to enable knee clearance and accommodate electrical wiring and interconnecting cabling.

The VDT screen should be adjustable to an angle of -5 to +15 degrees from the vertical plane at a viewing distance of 14 to 32 inches from the operator with the screen positioned 10 to 15 degrees below the operator's eye level. In addition, the screen should have brightness and contrast adjustment capability.

The keyboard should have a non-skid base, be non-reflective and detached from the screen. It is desirable that the keyboard be sloped when on a flat surface and positioned at a height (e.g., 28-31 inches) which enables the forearms to be horizontal and the upper arms vertical so that the elbows are at a right angle when using it. Use of wrist pads may be warranted for some personnel.

An adjustable document holder should be available to the user when there is a need to repeatedly refer from the screen to a document. Auxiliary lighting will often be necessary for proper document illumination.

The recommended illumination level for VDT work stations at which there is little, if any, need to read printed matter is 15-20 footcandles (i.e., about 150-200 lux) from overhead lighting. For work stations where reading of printed matter is required (e.g., an office environment) the illumination on the VDT screen (vertical plane) should be 15-20 footcandles from overhead lighting with 50-70 footcandles (i.e., about 500-700 lux) provided for the printed material by auxiliary lighting. In areas where overhead light creates glare on the screen, the installation of parabolic light fixtures can be provided to reduce the problem.

Noise levels in the work area should not exceed 55 dBA at 3 feet from equipment surfaces. In areas in which several printers may be operating simultaneously, it may be necessary to reduce noise levels by placing pads under the printers, enclosing them (either partially or totally) or replacing some with a quieter type.

The work environment temperature should be in the range of 68-77°F with a relative humidity between 40 and 60%. Ventilation of the work area should not create a draft on personnel and should be adequate to provide from 4 to 6 air changes per hour.

Special corrective lenses for eye glasses may be required by some personnel working at VDTs. This need can be determined on a case by case basis through a vision screening program.

REFERENCES

"Visual Display Units", Health and Safety Executive, ISBN 0 11 3685 4 (London: Her Majesty's Stationery Office, 1983).

McQuade, W., "Easing Tensions Between Man and Machines", Fortune, March 1984, pp. 58-96.

Smith, M., "Ergonomic Problems Dog VDT Use", Occupational Hazards, December 1987, pp. 39-41.

"Video Display Terminals - The Human Factor" (Chicago, IL: National Safety Council, 1982).

Brauninger, U., and E. Grandjean, "Lighting Characteristics of Visual Display Terminals From an Ergonomic Point of View", CHI'83, Proceedings of the Conference on Human Factors in Computing Systems, December 1983, pp. 274-276.

"Lighting up the CRT Screen - Problems and Solutions", Lighting Design & Application, January 1984, pp. 14-17.

Connell, J. J., "Managing Human Factors in the Automated Office", Modern Office Procedures, March 1982, pp. 1-10.

"Video Displays, Work and Vision" (Washington, D. C.: National Academy Press, 1983).

Cakir, A., et. al., "Visual Display Terminals" (New York: John Wiley & Sons Ltd., 1980).

WORKPLACE HEALTH PROTECTION PROGRAM ELEMENT CHECKLIST ON VISUAL DISPLAY TERMINALS

		Yes	No	Don't Know	N/A
1.	Have VDT users been trained to:				
	. Adjust chair?	___	___	___	___
	. Adjust screen position/ viewing angle?	___	___	___	___
	. Adjust lighting?	___	___	___	___
	. Avoid glare?	___	___	___	___
	. Take periodic rest breaks and stretching exercises?	___	___	___	___
2.	Is keyboard height acceptable (28-31")?	___	___	___	___
3.	Do VDT keyboards have a non-reflective surface?	___	___	___	___
4.	Are keyboards detachable from the VDT?	___	___	___	___
5.	Do keyboards have a non-skid base?	___	___	___	___
6.	Is the VDT screen adjustable:				
	. Vertically?	___	___	___	___
	. Horizontally?	___	___	___	___
	. To vary brightness/contrast?	___	___	___	___
7.	Is the screen viewing angle acceptable (-5° to 15° from the viewing plane)?	___	___	___	___
8.	Is the screen position satisfactory (i.e., 10°-15° below eye level)?	___	___	___	___
9.	Is the viewing distance between the screen and user satisfactory (14-32")?	___	___	___	___
10.	Is the VDT screen cleaned periodically?	___	___	___	___
11.	Is an adjustable, ergonomically designed swivel chair with 5 legs provided to VDT operators?	___	___	___	___
12.	Does the chair have a lumbar support?	___	___	___	___

N/A - Not Applicable

WORKPLACE HEALTH PROTECTION PROGRAM ELEMENT CHECKLIST ON VISUAL DISPLAY TERMINALS

		Yes	No	Don't Know	N/A
13.	Is chair height adequate to enable proper feet and leg positioning (feet on floor, thighs horizontal, knee at good angle - 90°-110°)?	___	___	___	___
14.	Are VDTs positioned so that glare is not a problem?	___	___	___	___
15.	Is there adequate space at VDT stations for all equipment/ accessories, etc.?	___	___	___	___
16.	Are there adequate electrical outlets available at the work station?	___	___	___	___
17.	Is there adequate room under the work table to accommodate equipment, wiring, legs?	___	___	___	___
18.	Is wiring/cabling under the work station orderly?	___	___	___	___
19.	Is work station height adjustable?	___	___	___	___
20	Is background color acceptable (non-reflective and not in stark contrast with the work station)?	___	___	___	___
21.	Is general illumination at VDT work stations adequate (15-20 footcandles)?	___	___	___	___
22.	Are VDT work areas uniformly lighted?	___	___	___	___
23.	Where appropriate, is there an adjustable document holder available and in use?	___	___	___	___

N/A - Not Applicable

WORKPLACE HEALTH PROTECTION PROGRAM ELEMENT CHECKLIST ON VISUAL DISPLAY TERMINALS

		Yes	No	Don't Know	N/A
24.	Is illumination on document holders adequate (50-70 footcandles)?	___	___	___	___
25.	Have sources of glare/reflection on the screen or in the field of view been eliminated:				
	. Overhead lighting?	___	___	___	___
	. Windows?	___	___	___	___
	. Auxiliary lighting?	___	___	___	___
26.	Are coverings (blinds/curtains) provided for windows?	___	___	___	___
27.	Are parabolic lighting fixtures provided, where needed?	___	___	___	___
28.	Is a polarized light filter available for the VDT screen, if needed?	___	___	___	___
29.	Are the following available to VDT operators, as needed:				
	. Wrist rests?	___	___	___	___
	. Arm rests?	___	___	___	___
	. Foot rests?	___	___	___	___
30.	Are VDT operators provided awareness training regarding the potential adverse health effects associated with VDT use?	___	___	___	___
31.	Are high volume VDT users (exceed three hours per day) encouraged to take rest breaks?	___	___	___	___
32.	Is the noise level in the work area acceptable (at or below 55 dBA)?	___	___	___	___
33.	Is the temperature and relative humidity of the work area acceptable?	___	___	___	___

N/A - Not Applicable

WORKPLACE HEALTH PROTECTION PROGRAM ELEMENT CHECKLIST ON VISUAL DISPLAY TERMINALS

	Yes	No	Don't Know	N/A
34. Is work area ventilation adequate (4-6 air changes per hour for an office type setting)?	___	___	___	___
35. Is air movement at the operator's position sufficiently low (less than 60 feet per minute) to prevent drafts?	___	___	___	___
36. Are complaints related to VDT health effect concerns promptly investigated?	___	___	___	___
37. Are appropriate specifications (related to health effect considerations) delineated when purchasing new VDT equipment and work station furniture?	___	___	___	___

N/A - Not Applicable

8. EMERGENCY PREPAREDNESS

WORKPLACE HEALTH PROTECTION PROGRAM ELEMENT
ON
EMERGENCY RESPONSE

WORKPLACE HEALTH PROTECTION PROGRAM ELEMENT ON EMERGENCY RESPONSE

OBJECTIVE

To provide information regarding the role of the industrial hygiene function in the planning of, and response to emergency situations in which release of a toxic material is possible, or has already occurred.

BACKGROUND

A number of acutely toxic materials are stored, used, handled, or generated in petroleum and petrochemical operations. If released, these materials have the potential to pose substantial risk to employees and the surrounding community. A list of these should be developed by each facility and emergency plans developed to minimize the adverse effects that could result from their release.

Some highly toxic substances may be present in large quantities. Since they are typically handled in closed systems, the likelihood for a release of a significant amount of toxic material is low, and health concerns due to high concentrations of these toxic materials/substances are unlikely to occur. There is, however, the potential for an accidental or catastrophic release. As a result, it is essential that petroleum and petrochemical facilities develop and implement appropriate disaster preparedness/response plans. These should include an analysis of the potential for a disaster to occur; the determination of potential effects and consequences of a catastrophic release; the identification of resources to be mobilized for responding to such an event; the development of a written plan to minimize adverse effects and return the facility to a normal state; and a program review system to assess plan adequacy and the identification of needs for improvement.

The resources needed in an emergency response situation include an effective communication system, responders, personnel to manage and coordinate activities, communications between in-house and off-site personnel, and individuals to make decisions as to the actions to take to minimize adverse health effects and eliminate the problem.

Everyone involved in an emergency response situation must have a defined role and be provided the appropriate training necessary to effectively carry out their responsibilities. During emergencies, management will typically rely on input from emergency responders on which to base their decisions and actions. It is essential that they receive adequate, accurate information.

CONSIDERATIONS

The industrial hygiene function should determine, in discussions with management, its role in the facility's emergency response program. If time permits, industrial hygiene personnel should seek an active role in emergency response. However, the industrial hygiene function can serve in an advisory role and participate in risk assessment and the development, implementation, testing, review and critique of the plan, as appropriate.

In the advisory role, the industrial hygiene function should have input into the selection of the personal protective equipment to be used by emergency responders and others. The industrial hygiene group could also provide advice on the monitoring equipment needed to assess the exposure risk associated with releases that may occur, as well as the calibration methods, maintenance requirements, and recordkeeping necessary to obtain quality monitoring data. Industrial hygiene should work with others to develop emergency exposure limits for various contaminants, particularly acute toxicants, to minimize risk to employees, residents in the surrounding community, livestock, wildlife, and the environment if a catastrophic release occurs.

Where time permits, industrial hygiene could consider a more active role. This may involve direct participation including: maintaining an accurate inventory of all chemicals (raw materials, products, etc.) and radioactive materials in the facility and their location; providing an up-to-date compilation of material safety data sheets for substances that are in the facility; being involved in the pre-incident hazard communication, responder training (on-site and off-site) and, when appropriate, the training of contractors and community residents; being the caretaker of the monitoring equipment to be used in emergency response situations; and be actively involved in the monitoring program to be carried out if an emergency situation arises as a result of the release of a toxic substance.

If responsibility for any, or all, of these activities is assigned to the industrial hygiene function, its role should be clearly defined and the group held accountable for its activities in emergency response situations.

REFERENCES

"Pocket Guide to Chemical Hazards, DHHS (NIOSH) Publication No. 90-117 (Cincinnati, OH: National Institute for Occupational Safety and Health, 1990).

"Emergency Response Guidebook", DOT Publication P 5800-4 (Chicago, IL: American Labelmark Company, 1987).

Kelly, R. B., "Industrial Emergency Preparedness" (New York: Van Nostrand Reinhold Company, 1989).

"Management of Process Hazards", API Recommended Practice 750 (Washington, D. C.: American Petroleum Institute, 1990).

Burgess, W. A., "Recognition of Health Hazards in Industry" (New York: John Wiley & Sons, Inc., 1981).

Johnson, J. S. and K. J. Anderson, Ed., "Chemical Protective Clothing", Volumes 1 and 2 (Akron, OH: American Industrial Hygiene Association, 1990).

Proctor, N. H., et al, "Chemical Hazards in the Workplace" (Philadelphia, PA: J. B. Lippincott Company, 1988).

WORKPLACE HEALTH PROTECTION PROGRAM ELEMENT CHECKLIST ON EMERGENCY RESPONSE

		Yes	No	Don't Know	N/A
1.	Is there a written disaster preparedness/response plan which addresses potential toxic releases in the facility?	___	___	___	___
2.	Are there alarms/signals to identify an emergency situation in the facility?	___	___	___	___
3.	Do all personnel in the facility know what to do if an emergency signal is activated?	___	___	___	___
4.	Are procedures in place for evacuating the handicapped if an emergency arises?	___	___	___	___
5.	Are drills carried out periodically to ensure emergency signals are audible/visible and personnel respond effectively?	___	___	___	___
6.	Are audits carried out periodically to determine the effectiveness of the total emergency response plan?	___	___	___	___
7.	Where appropriate, are wind socks positioned such that they can be seen throughout the facility to provide an indication of an appropriate evacuation route?	___	___	___	___
8.	Have personnel been informed how to use wind sock orientation as a basis for evacuation from, or accessing a release location?	___	___	___	___
9.	If an emergency situation occurs, do personnel know where to assemble so that an accounting can be made?	___	___	___	___
10.	Are responsibilities clearly defined for emergency responders?	___	___	___	___

N/A - Not Applicable

WORKPLACE HEALTH PROTECTION PROGRAM ELEMENT CHECKLIST ON EMERGENCY RESPONSE

		Yes	No	Don't Know	N/A
11.	Does the emergency response plan include assistance from outside agencies, (e.g., medical fire, etc.) as required, in toxic release situations?	___	___	___	___
12.	Do emergency response personnel know their responsibilities relative to their role if a toxic release occurs?	___	___	___	___
13.	Have emergency responders been adequately trained in first aid, rescue, use of personal protective equipment, and other emergency equipment?	___	___	___	___
14.	Have emergency response personnel been provided appropriate hazard communication training?	___	___	___	___
15.	Do emergency response personnel have appropriate protective equipment available for the contaminants/pathogens to which they may be exposed?	___	___	___	___
16.	Are the following factors considered in personal protective equipment selection:				
	. Concentration of contaminant?	___	___	___	___
	. Routes of exposure to contaminant?	___	___	___	___
	. Toxicity of material?	___	___	___	___
	. Effectiveness of personal protective equipment (e.g., protection factor)?	___	___	___	___
	. Possible length of exposure?	___	___	___	___
	. Nature of the task/job person will perform?	___	___	___	___
	. Comfort factors of wearer?	___	___	___	___
	. Fitness of the wearer?	___	___	___	___
	. Equipment fit?	___	___	___	___
	. Compatibility?	___	___	___	___

N/A - Not Applicable

WORKPLACE HEALTH PROTECTION PROGRAM ELEMENT CHECKLIST ON EMERGENCY RESPONSE

		Yes	No	Don't Know	N/A
17.	Are employee considerations a part of personal protective equipment selection:				
	. Preference?	___	___	___	___
	. Acceptability?	___	___	___	___
	. Comfort?	___	___	___	___
18.	Is personal protective equipment:				
	. Properly stored?	___	___	___	___
	. Inspected after usage?	___	___	___	___
	. Effectively cleaned?	___	___	___	___
	. Effectively maintained?	___	___	___	___
19.	Are personnel provided training on:				
	. Why protective equipment is being provided?	___	___	___	___
	. How to properly use personal protective equipment?	___	___	___	___
	. How to clean personal protective equipment?	___	___	___	___
	. How to maintain personal protective equipment?	___	___	___	___
20.	Do supervisory personnel enforce use of protective equipment when its use is indicated?	___	___	___	___
21.	Are the role and responsibilities of the industrial hygiene function clearly defined in the facility's emergency response plan?	___	___	___	___
22.	Does each individual in the industrial hygiene group know and understand their responsibility with respect to the emergency response plan?	___	___	___	___
23.	Under the facility's disaster plan, is responsibility defined for:				
	. Maintaining an inventory of materials (chemical, radiation sources, etc.) that are on site?	___	___	___	___

N/A - Not Applicable

WORKPLACE HEALTH PROTECTION PROGRAM ELEMENT CHECKLIST ON EMERGENCY RESPONSE

	Yes	No	Don't Know	N/A
. Maintaining and providing a complete, up-to-date compilation of MSDSs?	___	___	___	___
. Maintaining monitoring equipment for use in emergency response situations?	___	___	___	___
. Carrying out monitoring if an emergency develops (e.g., release of a hazardous substance)?	___	___	___	___
. Preparing and maintaining a list of the radioactive materials on site, and their location?	___	___	___	___
24. Does industrial hygiene provide support in:				
. Plan development?	___	___	___	___
. Plan implementation?	___	___	___	___
. Plan testing?	___	___	___	___
. Plan review?	___	___	___	___
. Emergency response critique?	___	___	___	___
. Emergency response hazard communication training?	___	___	___	___
. Monitoring instrument selection and use?	___	___	___	___
. Emergency response personal protective equipment selection?	___	___	___	___
. Developing exposure limits for emergency situations (applicable to employees and community residents)?	___	___	___	___
25. Is there adequate equipment available for monitoring during an emergency situation?	___	___	___	___
26. Is the monitoring equipment effectively maintained and calibrated?	___	___	___	___
27. Are monitoring equipment calibration records available?	___	___	___	___

N/A - Not Applicable

WORKPLACE HEALTH PROTECTION PROGRAM ELEMENT CHECKLIST ON EMERGENCY RESPONSE

	Yes	No	Don't Know	N/A
28. Are there an adequate number of trained personnel to carry out a toxic release monitoring program?	___	___	___	___
29. Has an emergency response monitoring strategy been developed for:				
. Risk assessment?	___	___	___	___
. Evacuation?	___	___	___	___
. Re-entry?	___	___	___	___
30. Are personnel adequately trained to use the monitoring equipment and interpret results?	___	___	___	___
31. Has management provided guidance on documentation of monitoring results if an emergency response situation arises?	___	___	___	___
32. Have off-site resources (e.g., industrial hygiene, safety, environmental, mutual aid, fire departments, etc.) that may be called upon, been provided training appropriate to respond to an emergency within or around the facility?	___	___	___	___
33. Could a neighbor's operations result in a need to evacuate your facility due to a hazardous material release at that (neighbor's) facility?	___	___	___	___
34. Is there established hazard communication between neighboring facilities regarding potential toxic release hazards?	___	___	___	___
35. Does the capability exist to monitor contaminants released by neighboring facilities?	___	___	___	___

N/A - Not Applicable

WORKPLACE HEALTH PROTECTION PROGRAM ELEMENT CHECKLIST ON EMERGENCY RESPONSE

		Yes	No	Don't Know	N/A
36.	Have contractors working on site been apprised of the facility's emergency response plans?	___	___	___	___
37.	Is there a readily available list of contractors who are on-site each day?	___	___	___	___
38.	Are contractors familiar with emergency evacuation plans and egress routes from the facility?	___	___	___	___
39.	Have potential high hazard areas and high risk facilities (e.g., schools, hospitals, etc.) been identified?	___	___	___	___
40.	Is there a meteorological station on site, or nearby, to obtain data for use in predicting areas of concern (re: high air concentrations of contaminant)?	___	___	___	___
41.	Are real time dispersion monitoring models used to predict downwind toxic and flammable vapor/gas cloud concentrations?	___	___	___	___

N/A - Not Applicable

9. MEDICAL SURVEILLANCE

WORKPLACE HEALTH PROTECTION PROGRAM ELEMENT
ON
AUDIOMETRIC TEST RESULTS

WORKPLACE HEALTH PROTECTION PROGRAM ELEMENT ON AUDIOMETRIC TEST RESULTS

OBJECTIVE

To provide information on the use of audiometric test results as a basis for identifying measures to improve the hearing conservation program, thereby reducing the incidence of noise induced hearing loss.

BACKGROUND

Audiometric test results provide a measure of the effectiveness of the hearing conservation program that has been developed, implemented, and maintained through the combined efforts of management, the medical department, the industrial hygiene function, and employees. When the hearing sensitivity of personnel is maintained throughout a period of employment, it is testament to the effectiveness of the hearing conservation program. If, however, audiometric test results show evidence of occupational hearing loss among personnel, then measures need to be implemented promptly to improve the program and prevent further noise induced hearing loss.

Factors which influence the course of noise induced hearing loss include the overall noise level in the work area; the frequency spectrum of the noise to which an employee is exposed; the time distribution of the individual's exposure; and susceptibility of the person to noise. Thus, control of hearing loss will depend upon adjustment of one or more of these factors, where possible and practical. For some situations, the use of hearing protection will be necessary to reduce hearing damage risk. To be effective, employees must be trained in the use of this personal protection and it must be the proper type for the noise exposure situation.

Information developed by the industrial hygiene function in carrying out area noise surveys and personal noise dosimetry studies is useful to the medical department in determining who to include in the hearing conservation program. Thus, areas/units/jobs for which there is a potential for employees to be exposed to an 8-hour time-weighted average noise level of 85 dBA or more should be identified to medical, so personnel working at these locations can be included in the audiometric test program.

Hearing loss which results from acute type noise exposure can also be detected by audiometric testing. This type of loss should be differentiated from hearing loss which results from excessive exposure to noise over a long period of time. Such information will facilitate implementation of appropriate control measures for preventing recurrences of the acute type hearing loss, as well as those that result from long-term exposure. Accomplishing this requires the combined effort of management, medical department personnel, the industrial hygiene function, and employees.

CONSIDERATIONS

Audiometric test results which show evidence of hearing loss should be discussed between medical department personnel, management, and the designated industrial hygienist. The purpose for such discussion is to identify and characterize the noise exposures the employee may have had in the course of his or her work assignments; determine if the loss is occupationally related; and develop an action plan to prevent further deterioration of the hearing sensitivity of the affected individual, as well as to minimize the potential for such an occurrence among personnel with the same job assignment, or working in the same area as the affected individual.

If an observed hearing loss is judged to have been due to occupational exposure to noise, the individual's exposure to noise should be reevaluated. Measures should be taken to ensure that available hearing protection is being used properly and is the appropriate type for the noise exposure. The industrial hygienist, or other person responsible for this program element, in cooperation with supervisory personnel, should retrain the affected person in the use and care of personal hearing protectors, ensure the person can use this personal protection properly, and, if necessary, provide a different type. If there are other similarly noise exposed personnel, they too should be retrained with respect to the use of personal hearing protection, as well as other aspects of the hearing conservation program.

Where appropriate, recommendations should be made to management for reducing noise exposure risk through the implementation of engineering and/or administrative controls. This may necessitate resurveys of work areas to identify noise sources which contribute significantly to the noise exposure.

REFERENCES

Anticagia, J. R., "Physiology of Hearing", Chapter 24, The Industrial Environment, its Evaluation and Control (Washington, D. C.: U.S. Department of Health, Education, and Welfare, 1972).

Michael, P. L., "Industrial Noise and Conservation of Hearing", Patty's Industrial Hygiene and Toxicology, Vol. 1, 3rd Edition, Chapter 10 (New York: John Wiley & Sons, Inc., 1978), pp. 275-357.

Schneider, E. J., et al, "The Progression of Hearing Loss from Industrial Noise Exposures", American Industrial Hygiene Association Monograph Series, 1981, pp. 81-89.

"Noise and Hearing Conservation" (Akron, OH: American Industrial Hygiene Association, 1981).

Royster, L. H ., et al, "Recommended Criteria for Evaluating the Effectiveness of Hearing Conservation Programs" (Akron, OH: American Industrial Hygiene Association Monograph Series, 1981,) pp. 111-119.

WORKPLACE HEALTH PROTECTION PROGRAM ELEMENT CHECKLIST ON AUDIOMETRIC TEST RESULTS

		Yes	No	Don't Know	N/A
1.	Are employees who should be included in the audiometric test program (based on noise survey results) effectively identified to medical?	___	___	___	___
2.	Are supervisory personnel aware of the noise exposure risk of personnel under their supervision?	___	___	___	___
3.	Has the hearing conservation program prevented occupationally related hearing loss among all noise exposed employees?	___	___	___	___
4.	Are there discussions between medical, management, and industrial hygiene personnel regarding hearing loss cases?	___	___	___	___
5.	If observed, is an effort made to determine whether a hearing loss is due to occupational noise exposure?	___	___	___	___
6.	Are supervisory personnel included in discussions when personnel under their supervision experience loss of hearing?	___	___	___	___
7.	Is an action plan in place, for follow-up on hearing loss to ensure:				
	. Exposure risk has not changed?	___	___	___	___
	. Supervisory personnel are aware of the noise exposure risk and require use of hearing protection?	___	___	___	___
	. Measures are being taken to reduce employee exposure to noise?	___	___	___	___
	. Personnel are aware of the hearing damage risk potential associated with their job assignment?	___	___	___	___
	. Personnel are using personal hearing protection properly?	___	___	___	___
	. Personal hearing protection is the appropriate type for the noise exposure?	___	___	___	___

N/A - Not Applicable

WORKPLACE HEALTH PROTECTION PROGRAM ELEMENT CHECKLIST ON AUDIOMETRIC TEST RESULTS

	Yes	No	Don't Know	N/A
. Affected personnel are retrained with respect to their use of personal protection and the elements of the hearing conservation program?	___	___	___	___
8. Does evidence of hearing loss among employees result in a more concerted effort to reduce employee exposure to noise through:				
. Increased hazard awareness training?	___	___	___	___
. Better enforcement in the use of hearing protectors?	___	___	___	___
. Implementation of administrative controls?	___	___	___	___
. Installation of engineering controls?	___	___	___	___
9. Are periodic area noise and noise dosimeter surveys conducted to assess changes in noise exposure?	___	___	___	___
10. Is noise survey equipment periodically calibrated by a vendor?	___	___	___	___
11. Is personal hearing protection selected based on its suitability for a given noise environment?	___	___	___	___
12. Has there been an assessment of the noise hazard awareness aspect of the hearing conservation program to ensure its effectiveness?	___	___	___	___
13. Does medical counsel employees on the results of each audiometric test?	___	___	___	___
14. Is there documentation that personnel have been notified of the results of their hearing tests?	___	___	___	___
15. Have audiometric test procedures been reviewed to ensure adherence with recommended practices.	___	___	___	___

N/A - Not Applicable

WORKPLACE HEALTH PROTECTION PROGRAM ELEMENT CHECKLIST ON AUDIOMETRIC TEST RESULTS

		Yes	No	Don't Know	N/A
16.	Is there documentation that the audiometric testing environment meets established standards?	___	___	___	___
17.	Is there adequate documentation that the audiometer is functioning properly?	___	___	___	___
18.	Are audiometric test records adequate for effective identification of the incidence of hearing loss among personnel?	___	___	___	___

N/A - Not Applicable

WORKPLACE HEALTH PROTECTION PROGRAM ELEMENT
ON
BIOLOGICAL TEST RESULTS

WORKPLACE HEALTH PROTECTION PROGRAM ELEMENT ON BIOLOGICAL TEST RESULTS

OBJECTIVE

To identify the need for biological monitoring to substantiate air monitoring results and the need for medical, industrial hygiene, and management to jointly investigate situations where Biological Exposure Indices (BEIs) are exceeded.

BACKGROUND

Biological monitoring, which is performed by the medical department, is a method to assess workers' exposure to chemicals that may be present in a place of employment. This is accomplished through the measurement of appropriate determinants in biological specimens collected from the worker at a specified time. The determinant, which indicates the presence of the substance of interest in the body, can be the material itself, its metabolite, or a characteristic reversible biochemical change induced by the agent. Such determinations may be made in exhaled air, urine, blood, or other biological specimen collected from the exposed worker. The biological monitoring result may indicate the intensity of a recent exposure, the average daily exposure, or the cumulative exposure of an individual, depending upon the specimen obtained, the determinant selected, and/or the time elapsed since the exposure occurred.

Biological Exposure Indices (BEIs) have been published by the American Conference of Governmental Industrial Hygienists. They represent the levels of determinants that are most likely to be observed in biological samples obtained from a healthy worker who has received an inhalation exposure to the 8-hour Threshold Limit Value (TLV) for that substance. The BEIs are not values which, if exceeded, indicate that previous hazardous exposures have occurred, but rather, are an indication that the job assignment should be investigated with a view toward reducing exposure. Such investigation is typically carried out by the industrial hygiene function and the employee's supervisor at the request of the medical department. In some instances, medical department personnel participate in the investigation.

BEIs have been developed for a number of contaminants to which personnel are potentially exposed in the petroleum and petrochemical industries. The dose received by an individual exposed to these substances can be determined in a biological monitoring program and the results compared to the appropriate index to determine if there is an exposure concern for a specific job hazard. The result provides an indication of total dose received (i.e., the amount absorbed by inhalation, dermal absorption, and ingestion).

Biological monitoring may be conducted to substantiate air monitoring results, test the effectiveness of personal protective equipment, determine exposure via skin absorption and ingestion, or provide an indication of non-occupational exposure to a substance. Thus, this procedure is complementary to air monitoring, and should be carried out when it offers advantages to the air monitoring program.

If a biological monitoring result indicates that a BEI determinant level has been exceeded, it would be advisable to repeat the test to substantiate the result, and review the worker's job assignment to ensure that appropriate procedures/work practices are being followed and that an exposure hazard does not exist. The latter will require communication between the medical department and industrial hygiene personnel with involvement of the employee's supervisor in the conduct of the job/task review and exposure assessment.

CONSIDERATIONS

When practical, a biological monitoring program could be considered for locations where there is potential exposure to substances for which BEIs have been established. This effort would be in addition to the medical surveillance requirements specified in applicable regulations and surveillance programs that the medical department has implemented. In each situation, the employee should be informed of the results and the findings made a part of the individual's medical record.

If it has been satisfactorily demonstrated that a BEI has been exceeded, an investigation should be carried out as a joint effort by management, medical, and industrial hygiene personnel. The findings of such investigations should be documented, along with information on the measures that have been implemented to reduce worker exposures to the agent of concern, as well as the results of follow-up biological tests which demonstrate that the dose received by the exposed individuals(s) has been effectively reduced. When considered beneficial, the findings of such investigations could be communicated to other facilities where similar exposure situations may exist. This will enable those locations to implement measures to prevent employees from receiving an excessive dose of the contaminant.

REFERENCES

"Cooper, W. C., "Health Surveillance Programs in Industry", Patty's Industrial Hygiene and Toxicology, Vol. III (New York: John Wiley & Sons, Inc., 1979), pp. 595-609.

Soule, R. D. "Sampling and Analysis", Patty's Industrial Hygiene and Toxicology, Vol. 1, 3rd Edition (New York: John Wiley & Sons, Inc., 1978), pp. 762-768.

"Threshold Limit Values for Chemical Substances and Physical Agents and Biological Exposure Indices", Latest Edition (Cincinnati, OH: American Conference of Governmental Industrial Hygienists).

Fiserova-Bergerova, V., "Development of Biological Exposure Indices (BEIs) and Their Implementation", Applied Industrial Hygiene Journal, March 1987, pp. 87-92.

Thomas, A. A., "Skin Absorption: A Potential Contributing Factor to the BEI", Applied Industrial Hygiene Journal, July 1986, pp. 87-90.

WORKPLACE HEALTH PROTECTION PROGRAM ELEMENT CHECKLIST ON BIOLOGICAL TEST RESULTS

		Yes	No	Don't Know	N/A
1.	Has a determination been made of the need to implement a biological monitoring program?	___	___	___	___
2.	Do the biological tests that are carried out on personnel include:				
	. Periodic tests for specific determinants?	___	___	___	___
	. Those required by regulation?	___	___	___	___
	. On an ad hoc basis?	___	___	___	___
3.	Are biological tests carried out to substantiate an excessive inhalation exposure to a specific contaminant?	___	___	___	___
4.	Are biological tests carried out to determine skin absorption as a route of exposure?	___	___	___	___
5.	Is biological monitoring performed for determining exposure to:				
	. Benzene?	___	___	___	___
	. Carbon Monoxide?	___	___	___	___
	. Ethyl benzene?	___	___	___	___
	. Fluorides?	___	___	___	___
	. Furfural?	___	___	___	___
	. n-Hexane?	___	___	___	___
	. Lead:				
	- Organic?	___	___	___	___
	- Inorganic?	___	___	___	___
	. Phenol?	___	___	___	___
	. Ionizing radiation?	___	___	___	___
	. Toluene?	___	___	___	___
	. Xylenes?	___	___	___	___
	. Other?(__________________)	___	___	___	___
6.	Have biological monitoring results always been below applicable Biological Exposure Indices?	___	___	___	___
7.	Is there effective communication between medical and industrial hygiene on the scope of the biological monitoring program and the results of tests?	___	___	___	___
8.	Are investigations carried out to determine the cause of high biological monitoring results?	___	___	___	___

N/A - Not Applicable

WORKPLACE HEALTH PROTECTION PROGRAM ELEMENT CHECKLIST ON BIOLOGICAL TEST RESULTS

	Yes	No	Don't Know	N/A
9. Do the following participate in investigations to determine the cause for a high biological monitoring result:				
. Medical?	___	___	___	___
. Management?	___	___	___	___
. Industrial hygiene?	___	___	___	___
. Other? (________________)				
10. Are findings of the investigations to determine the cause for high biological monitoring results effectively documented?	___	___	___	___

N/A - Not Applicable